AF539220

Principles of Human Nutrition

Principles of Human Nutrition

Shivani Rawat

RANDOM PUBLICATIONS
NEW DELHI (INDIA)

Principles of Human Nutrition

ISBN 978-93-5111-523-6

Published in 2015 in India by

RANDOM PUBLICATIONS

4376-A/4B, Gali Murari Lal, Ansari Road
New Delhi-110 002
Phone : +9111-43580356, 011-23289044, 011-43142548
e-mail: sales@randompublications.com,
info@randompublications.com, randomexports@gmail.com

Type Setting by : Friends Media, Delhi-110089
Printed at : Sanat Printers

Preface

Human nutrition is the provision to obtain the materials necessary to support life. In general, people can survive for two to eight weeks without food, depending on stored body fat and muscle mass. Survival without water is usually limited to three or four days. Lack of food remains a serious problem, with about 36 million people dying every year from causes directly or indirectly related to hunger. Childhood malnutrition is also common and contributes to the global burden of disease. However global food distribution is not equal, and obesity among some human populations has increased to almost epidemic proportions, leading to health complications and increased mortality in some developed, and a few developing countries. Obesity is caused by consuming more calories than are expended, with many attributing excessive weight gain to a combination of overeating of "unhealthy" (high fat, high sugar, high carbohydrate) foods and insufficient exercise.

Nutritional science investigates the metabolic and physiological responses of the body to diet. With advances in the fields of molecular biology, biochemistry, and genetics, the study of nutrition is increasingly concerned with metabolism and metabolic pathways: the sequences of biochemical steps through which substances in living things change from one form to another.

This book is written as a course text for those studying degree courses in nutrition and dietetics and for students on modular courses on nutrition within other degree courses, e.g. food studies, medicine, health sciences, nursing and biological sciences. It is also of great value as a reference for professional nutritionists and dietitians, food scientists and health professionals based in academia, in practice and in commercial positions such as within the food and pharmaceutical industries.

Author

Contents

Preface v

1. Importance of Nutrition **1**

What is Nutrition? 2

Significance of Good Nutrition 5

Nutrition and Medicine 7

Nutritional Status 8

Relation between Food and Nutrition 10

Food Grains Development 11

Breeding for Nutritional Quality 16

Food-Energy Systems 18

Food-Energy Nexus 28

2. Basic Human Nutrients **34**

Classes of Nutrients 36

Carbohydrates 37

Fibers 37

Fat 38

Protein 40

Minerals 41

Vitamins 43

Water 43

Other Nutrients 45

Advice and Guidance 47

Healthy Diets 49

3. Healthy Diet **52**

Nutrient Dense Foods 53

Components of Healthy Diet 57

Tips for Dietary Recommendations in Daily Routine 72

Variety of Foods in Diet 77

Nutritional Tools and Guidelines for Planning Healthy Diets 84

Measurement of Nutritional State 86

4. Carbohydrates and Nutrition **98**

Classification 100

Carbohydrates in Nutrition 101

Analysis of Dietary Carbohydrate 105

Intake of Carbohydrates 108

Physiological Effects of Carbohydrates 111

Carbohydrates in Health Maintenance 117

Carbohydrate and Disease 122

Role of the Glycemic Index in Food Choices 128

Guidelines for Carbohydrate Food Choices 132

5. Energy and Protein Requirements **137**

Individuals and Groups 139

Common Considerations to the Estimation of Energy and Protein Requirements 139

Principles for the Estimation of Energy Requirements 150

Principles of Estimating Protein Requirements 161

The Protein-Energy Ratio as a Measure of Dietary Quality 174

6. Role of Dietary Fats and Oils **181**

Composition of Dietary Fat 181

Digestion and Metabolism 186

Availability of Edible Fats and Oils 193

Processing and Refining Edible Oils 197

Selected Uses of Fats and Oils in Food 208

7. Causes of Malnutrition **212**

The Scale of the Problem 213

Nutrition Improvement: Nature and Evolution 215

A Framework for Causes of Malnutrition 218

Promotion and Protection of Nutritional Well-being 219

Food Production and Food Security 224

Nutrition and Infection, Health and Disease 233

8. Socio-cultural Aspects of Human Nutrition **245**

Food Habits 245

Nutritional Advantages of Traditional Food Habits 247

Food Taboos 248

Changing Food Habits 250

Harmful New Habits 251

Influencing Change for the Better 252

Population, Food and Nutrition 253

Reproduction and Nutritional Status 257

Breastfeeding, Fertility and Family Planning 258

Nutrition During Particular Times in the Life Cycle 260

Breastfeeding 273

Bibliography **288**

Index **290**

1

Importance of Nutrition

The importance of good nutrition is nothing new. Back in 400 B.C., Hippocrates said, "Let food be your medicine and medicine be your food." Today, good nutrition is more important than ever. At least 4 of the 10 leading causes of death— heart disease, cancer, stroke and diabetes— are directly related to the way we eat; diet is also implicated in scores of other conditions. But while the wrong diet can be deadly, eating right is among the cornerstones of health.

Of course, food alone isn't the key to a longer and healthier life. Good nutrition should be part of an overall healthy lifestyle, which also includes regular exercise, not smoking or drinking alcohol excessively, stress management and limiting exposure to environmental hazards. And no matter how well you eat, your genes play a big part in your risk for certain health problems. But don't underestimate the influence of how and what you eat.

For example, atherosclerosis (hardening of the arteries) can begin in early childhood, but the process can be halted — even reversed — if you make healthy changes in your diet and lifestyle. The gradual bone thinning that results in osteoporosis may be slowed if you consume enough calcium, maintain adequate Vitamin D levels and participate in weight-bearing exercise. You may be genetically predisposed to diabetes, but keep your weight within a healthy range through diet and exercise and the disease may never strike you.

The keys to good nutrition are balance, variety and moderation. To stay healthy, your body needs the right balance of carbohydrates, fats, and protein — the three main components of nutrition.

You also need vitamins, minerals and other substances from many different foods, and while some foods are better than others, no single food or food group has it all — so eating a variety of different foods is essential.

Moderation means eating neither too much nor too little of any food or nutrient. Too much food can result in excess weight and even too much of certain nutrients, while eating too little can lead to numerous nutrient deficiencies and low body mass.

What is Nutrition?

Nutrition is the study of the relationship between diet and states of health and disease. Health can be brought about by an imbalance of nutrients, producing either an excess or deficiency which in turn affects body functioning in a cumulative manner. Moreover, because most nutrients are, in some way or the other, involved in cell-to-cell signalling, deficiency or excess of various nutrients affects hormonal function also indirectly. Thus, because they largely regulate the expression of genes, hormones represent a link between nutrition and how our genes are expressed, i.e. our phenotype.'

The strength and nature of this link are continually under investigation, but observations especially in recent years have demonstrated a pivotal role for nutrition in hormonal activity and function and therefore in health. Mineral and/or vitamin (tocotrienol and tocopherol) deficiency or excess may yield symptoms of diminishing health such as goitre, scurvy, osteoporosis, weak immune system, disorders of cell metabolism, certain forms of cancer, symptoms of premature aging, and poor psychological health.

As of 2005, twelve vitamins and about the same number of minerals are recognised as 'essential nutrients', meaning that they must be consumed and absorbed—or, in the case of vitamin D, alternatively synthesised via UVB radiation—to prevent deficiency symptoms and death. Certain vitamin-like substances found in foods, such as carnitine, have also been found essential to survival and health, but these are not strictly 'essential' to eat because the body can produce them from other compounds.

Moreover, thousands of different phytochemicals have recently been discovered in food (particularly in fresh vegetables), which have many discovered and yet to be discovered properties including antioxidant activity. Other essential nutrients include essential amino acids, choline and the essential fatty acids.

In addition to sufficient intake, an appropriate balance of essential fatty acids—omega-3 and omega-6 fatty acids—has been discovered to be crucial for maintaining health. Both of these unique "omega" long-chain polyunsaturated fatty acids are substrates for a class of eicosanoids known as prostaglandins. The omega-3 eicosapentaenoic acid (EPA) (which can be made in the body from the omega-3 essential fatty acid alpha-linolenic acid (LNA), or taken in through marine food sources), serves as building block for series 3 prostaglandins (e.g. weakly-inflammation PGE3).

The omega-6 dihomo-gamma-linolenic acid (DGLA) serves as building block for series 1 prostaglandins (e.g. anti-inflammatory PGE1), whereas arachidonic acid (AA) serves as building block for series 2 prostaglandins (e.g. pro-inflammatory PGE1). Both DGLA and AA are made from the omega-6 linoleic acid (LA) in the body, or can be taken in directly through food.

An appropriately balanced intake of omega-3 and omega-6 partly determines the relative production of different prostaglandins, which partly explains the importance of omega-3/omega-6 balance for cardiovascular health. In industrialised societies, people generally consume large amounts of processed vegetable oils that have reduced amounts of essential fatty acids along with an excessive amount of omega-6 relative to omega-3.

The rate of conversions of omega-6 DGLA to AA largely determines the production of the respective prostaglandins PGE1 and PGE2. Omega-3 EPA prevents AA from being released from membranes, thereby skewing prostaglandin balance away from pro-inflammatory PGE2 made from AA toward anti-inflammatory PGE1 made from DGLA.

Moreover, the conversion (desaturation) of DGLA to AA is controlled by the enzyme delta-5-desaturase, which in turn is controlled by hormones such as insulin (up-regulation) and glucagon (down-regulation). Because different types and amounts of food eaten/absorbed affect insulin, glucagon and other hormones to varying degrees, not only the amount of omega-3 versus omega-6 eaten but also the general composition of the diet therefore determine health implications in relation to essential fatty acids, inflammation (e.g. immune function) and mitosis.

Several lines of evidence indicate lifestyle-induced hyperinsulinemia and reduced insulin function (i.e. insulin resistance) as a decisive factor in many disease states. For example, hyperinsulinemia and insulin resistance are strongly linked to chronic inflammation, which in turn is strongly linked to a variety of adverse developments such as arterial microinjuries and clot formation (i.e. heart disease) and exaggerated cell division (i.e. cancer).

Hyperinsulinemia and insulin resistance (the so-called metabolic syndrome) are characterised by a combination of abdominal obesity, elevated blood sugar, elevated blood pressure, elevated blood triglycerides, and reduced HDL cholesterol. The negative impact of hyperinsulinemia on prostaglandin PGE1/PGE2 balance may be significant.

The state of obesity clearly contributes to insulin resistance, which in turn can cause type 2 diabetes. Virtually all obese and most type 2 diabetic individuals have marked insulin resistance. Although the association between overfatness and insulin resistance is clear, the exact (likely multifarious) causes of insulin resistance remain less clear.

Importantly, it has been demonstrated that appropriate exercise, more regular food intake and reducing glycemic load all can reverse insulin resistance in overfat individuals (and thereby lower blood sugar levels in those who have type 2 diabetes). Overfatness can unfavourably alter hormonal and metabolic status via resistance to the hormone leptin, and a vicious cycle may occur in which insulin/leptin resistance and overfatness aggravate one another. The vicious cycle is putatively fuelled by continuously high insulin/leptin stimulation and fat storage, as a result of high intake of strongly insulin/leptin stimulating foods and energy.

Both insulin and leptin normally function as satiety signals to the hypothalamus in the brain; however, insulin/leptin resistance may reduce this signal and therefore allow continued overfeeding despite large bodyfat stores. In addition, reduced leptin signalling to the brain may reduce leptin's normal effect to maintain an appropriately high metabolic rate.

There is debate about how and to what extent different dietary factors—e.g. intake of processed carbohydrates, total protein, fat, and carbohydrate intake, intake of saturated and trans fatty acids, and low intake of vitamins/minerals—contribute to the development of insulin—and leptin resistance. In any case, analogous to the way modern man-made pollution may potentially overwhelm the environment's ability to maintain 'homeostasis', the recent explosive introduction of high Glycemic Index—and processed foods into the human diet may potentially overwhelm the body's ability to maintain homeostasis and health,

Antioxidants are another recent discovery. As cellular metabolism/ energy production requires oxygen, potentially damaging (e.g. mutation causing) compounds known as radical oxygen species or free radicals may form. For normal cellular maintenance, growth, and division, these free radicals must be sufficiently neutralised by antioxidant compounds, such as certain vitamins (vitamin C, vitamin E, vitamin K and the aforementioned phytochemicals as well as other compounds, some of which the body itself produces. Different antioxidants are now known to function in a cooperative network, e.g. vitamin C can reactivate free radical-containing glutathione or vitamin E by accepting the free radical itself, and so on.

It is now also known that the human digestion system contains a population of a range of bacteria which are essential to digestion, and which are also affected by the food we eat. The role and significance of the intestinal bacterial flora is under investigation.

Nutrition is very important for improving sports performance. The most common means to improve performance through diet is the practice of eating large quantities of protein, usually red meat, when attempting to build muscle mass; its efficacy is doubtful, as daily protein intake even on a normal diet usually outweighs the amount of muscle protein which can be synthesised in a day, and protein is a much less efficient source of the energy needed to build new muscle tissue than are fats and carbohydrates.

Significance of Good Nutrition

Most governments provide guidance on good nutrition, and some also impose mandatory labelling requirements upon processed food manufacturers to assist consumers in complying with such guidance. Current dietary guidelines in the United States are presented in the concept of a food pyramid. There is no apparent consisteny in science-based nutritional recommendations between countries, indicating the role of politics as well as cultural bias in research emphasis and interpretation.

Nutrition is everyone's challenge, but no one's sole responsibility. To help target audience understand their potential contributions, the Task Force identified five types of actions that can lead to optimal nutrition for women and children, including:

— Personal actions such as food choices, food safety practices, and physical activity

— Community actions such as the availability of nutritious and safe foods, outreach services and educational opportunities

— Health care provider actions such as the reinforcement of community health messages and the acquisition and use of knowledge related to the life cycle perspective and nutritional influences on health

— Regional actions. An action-oriented agenda and educational tools; technical assistance in developing community-specific food-based dietary guidelines and intervention strategies

— National and international actions such as the development of nutrition and food safety policies; infrastructure to provide a nutritionally adequate and safe food supply, education, supplementation and/or fortification if necessary; and the implementation of appropriate world health resolutions and codes

Importance of Lifecycle

A life cycle perspective, in physiological terms, is an important public health concept because conditions early in life-even before birth-may affect an individual's health, growth, and development over his or her life span. Similarly, certain health conditions may affect the health and growth of a woman's offspring as well.

Women's nutritional status before conception may contribute to positive or negative outcomes during pregnancy and in the infant. For example:

— Women with low folate status in the periconceptional period are at significantly elevated risk of giving birth to a child with spina bifida or a related neurological defect.

— Women who are underweight before pregnancy, particularly immature adolescents, have a higher risk of giving birth to a low-birthweight infant.

— Women who are obese before conception may experience complications during pregnancy and childbirth. In addition, they are at higher risk of having babies with congenital malformations.

In addition, many of the health problems faced by women and their young children are interrelated, such as the following:

— Vitamin and mineral deficiencies and low birthweight contribute to the impaired growth and development of infants and young children.

— Mothers who were low-birthweight at birth are at increased risk of delivering a low birthweight infant.

— Mothers who are overweight/obese prior to conception or as a result of excessive weight retention after pregnancy are at an increased risk for a variety of diseases later in life.

Good nutrition at each of these stages of the life cycle contributes to optimal growth and development of infants and children and a healthier adult population. The Task Force fully recognises that the nutrition of older children is important, as is the nutrition of adolescent boys and of men. Nonetheless, it is the women, infants, and very young children who are most vulnerable. Attending to their nutrition is most crucial for ensuring a healthy start.

Nutrition and Medicine

Nutrition influences clinical practice in all branches of medicine and is important in all stages of life. Clinical nutrition is the application of the principles of nutrition science and medical practice to the diagnosis, treatment, and prevention of human disease caused by the deficiency, excess, or metabolic imbalance of nutrients.

Malnutrition and other deficiency states, such as marasmus, kwashiorkor, xerophthalmia, and rickets, are major causes of morbidity and mortality not only in developing countries but also in industrialised countries under conditions of deprivation. Malnutrition occurs in alcohol and drug addiction, in prolonged illnesses of various etiologies, and as a complication of some surgical and medical procedures.

Nutrition is compromised in many systemic diseases, sometimes with serious effects on health outcome. Thus, many medical centers have established multidisciplinary nutrition support teams including physicians, surgeons, nurses, dietitians, pharmacists, and laboratory technicians. Such a team identifies patients requiring nutritional support, determines their nutritional status, recommends therapeutic diets, and provides long-term follow-up. If an adequate intake cannot be maintained orally, enteral or parenteral nutrition is given as required.

Nutritional factors may play a role in the causation of several chronic degenerative diseases, such as cancer, hypertension, and coronary artery disease. Special diets are important in the management of many hereditary metabolic disorders, such as galactosemia and phenylketonuria.

Nutritional Status

The appraisal of nutritional status should be a part of every general health evaluation, including history, physical examination, and selected laboratory tests.. Pertinent issues may be rate of growth and development in infants and children, body composition in children and adults, or evidence of specific deficiencies or excesses of essential nutrients in any patient.

The nutritional history is inevitably intertwined with the medical history, which can often provide clues to the nature of nutritional disease. For instance, a history of GI bleeding can account for iron deficiency anemia; treatment of acne with vitamin A can lead to vitamin A toxicity manifested by headache, nausea, and diplopia; and an enlarged thyroid gland may be due to iodine deficiency. Conditions that predispose to nutritional disease include gross underweight, gross overweight, recent weight loss, alcoholism, malabsorption, hyperthyroidism, protracted fever, sepsis, fad diets, drug usage, and psychiatric disorders.

Diet history taking should include a 24-h recall of foods eaten and a food frequency questionnaire that asks about foods or food groups frequently consumed. More detailed information may be obtained from a food diary, in which the patient records what is eaten during a 3-day period, or a weighed ad libitum diet, in which food is selected by the patient and weighed each day for 3 days to a week. The last method is the most accurate and is usually reserved for clinical research.

The physical examination can provide clues to nutritional disease. Any body system can be affected by a nutritional disease. For example, the CNS is affected in pellagra, beriberi, pyridoxine deficiency or excess, and vitamin B12 deficiency. Taste and smell are affected by zinc deficiency. Hypertension, diabetes, and coronary artery disease are associated with obesity.

The GI system can be injured by malnutrition and alcoholism. The oral cavity (lips, tongue, teeth, gums, and buccal mucosa) is affected by vitamin B-complex deficiency and scurvy. The effects of malnutrition on the skin can include rashes, petechial hemorrhages, ecchymoses, pigmentation, edema, and dryness. Bones and joints are diseased in rickets, osteomalacia, osteoporosis, and scurvy.

Imaging procedures and biochemical laboratory tests are useful in appraising nutritional status. X-rays of the thorax and the skeleton are used to determine cardiopulmonary function and bone density. GI disturbances secondary to malnutrition can be studied radiographically with contrast media. CT and MRI are useful in viewing soft tissues.

Modern analytical instrumentation with high-pressure liquid chromatography, radioimmunoassay of enzymes, or flame photometry has greatly improved the sensitivity and specificity of biochemical tests of nutritional status. Measuring plasma levels or urinary excretion of proteins, lipids, electrolytes, trace minerals, and vitamins can provide evidence about body stores of these nutrients.

Nutrient-dependent enzymatic tests can be applied to both red and white blood cells, and the immune status can be appraised by determining lymphocyte counts, immunoglobulin levels, and the response of lymphocytes to mitogens and by performing skin tests.

Role of Diet Quality

Vitamin and Mineral Adequacy: Micronutrient undernutrition can impact both the length and quality of life by increasing the risk of morbidity and mortality. Problems of deficiencies among women and very young children are best documented for seven micronutrients: iron, iodine, vitamin A, zinc, folate, vitamin D and calcium. These vitamins and minerals enable us to use the energy provided by foods; build, maintain and repair cell and tissue structures; and perform critical biochemical transformations.

An adequate intake of vitamins and minerals is particularly important for adolescent girls in preparation for motherhood, during the childbearing years of all women (above all while they are pregnant and lactating), and for infants and children for at least the first two years after birth. During these critical periods, diet quality can have far-reaching effects on growth (including stature), development (including brain functions and learning ability), and immune responses to infections and disease.

Global estimates of the prevalence of micronutrient deficiencies vary by nutrient. The most reliable, quantitative estimates of deficiencies indicate that over 700 million people are affected by iodine deficiency disorders, 250 million children to five years of age are affected by and at risk for vitamin

A deficiency, and roughly 40% of children to five years of age in developing countries and women worldwide are iron-deficient or anemic.

During the childbearing years, a woman's weight can affect outcomes of pregnancy and have far-reaching effects on the woman's health as well. Women who are very thin or very short when they conceive are at increased risk for delivering an infant who has fetal growth restriction. This condition contributes strongly to the high prevalence of low birthweight (weight less than 2,500 g) in the developing world.

About four million infants born this small die. Many of the surviving infants are at increased risk for cognitive and neurological deficits and other adverse health outcomes. In addition, low-birthweight infants are more likely to become small adults. This may reduce their work capacity, and among women, it increases the risk of delivering a low birthweight infant.

The prevalence of overweight and obesity is alarmingly high in many countries and increasing rapidly throughout much of the world-in industrialised and developing nations alike. Currently, for example, more than 40% of nonpregnant U.S. women ages 15 to 49 years are overweight or obese.

The majority of girls and adolescents who are overweight or obese will become overweight or obese adults. Obesity before pregnancy increases women's risk during the entire life cycle. For example, obesity increases risks for infertility, maternal and fetal complications during pregnancy (including congenital malformations) and delivery. During the postpartum period, obesity may impair lactation performance and, later in life, increase risk of chronic diseases such as cardiovascular disease and Type II diabetes.

Infants and Children Under Two Years: In developing countries, stunting (short for age) affects up to 64% of children less than five years of age, underweight (weight low for age) affects up to 56%, and wasting (weight low for length) affects up to 25%. Undernutrition, inappropriate breastfeeding practices, and inappropriate complementary feeding practices contribute significantly to most cases of impaired growth. Children with impaired growth resulting from undernutrition are at increased risk of infection, cognitive deficits, and even death. Like low-birthweight infants, they are more likely to become small adults.

Relation between Food and Nutrition

The last 30 years have witnessed spectacular increases in food-grain production in India, from 51 million tonnes to more than 130 million tonnes. A sizeable buffer stock has also been built up to face the likely shortages arising out of uncertain production levels Various national projections indicate that food-grain production must be doubled to meet the needs of an estimated population of more than 950 million by the turn of this century. Since food grains constitute the principal source of calories, proteins, and other nutrients (table1), careful efforts are necessary not only to conserve them against qualitative and quantitative losses but also to upgrade their nutritional and acceptability features. Thus, a systems approach has become crucial for integrating the production, conservation, and nutrition of food grains to get maximum benefits from national efforts.

Table 1. Daily Per Capita Food Consumption and Recommended Levels Recommended level (g)

Food materials	Consumption (g)	ICMR	Task Force on
Cereals	434	412	376
Pulses	34	60	64
Green leafy vegetables	21	125	116
Other vegetables	71	50	61
Roots and tubers	—	87	69
Fruits	10	30	40
Milk	69	100	189
Fats/oils	12	40	39
Meat and fish (and egg)	14	30	23
Eggs	—	30	15
Sugar/jaggery	19	35	38

Breeding of new food-grain varieties has been directed to increasing per hectare yields and resistance against field-borne micro-organisms and insect pests. Advances in food technology and nutrition have, however, given some insight into the desirable features that need to be considered in breeding programmes.

Food Grains Development

About 70 per cent of food grains produced in India are retained for farm-level consumption and the rest moves along a chain of agencies before it reaches the consumption points. Post-harvest conservation by modern procedures is therefore a crucial need to prevent the dissipation of national efforts to raise food production levels. The incidence of bunt in wheat, chalky grains in rice, and Gibberella infection in maize, and the impairment of processing qualities as a result of pre-harvest infection have engaged the attention of scientists in recent years.

The expertise in food conservation built up during the last 30 years has found increasing application, but basic information to evolve varieties with desirable storage, processing, and nutritional or organoleptic qualities is important in meeting future needs. Variable production levels in different years emphasise the need for varieties that give maximum yields during processing and suffer minimum losses during post-harvest handling and storage.

Food grains harvested from standing crops are dried and processed before cooking and consumption. Milling is used to obtain rice from paddy, dhals from pulses, and flours from wheat and millets. The wastages inherent in some traditional practices have been minimised through improved processing technology and equipment design. The physico-chemical features of food grains need attention in varietal development programmes.

About 50 per cent of all cereal production consists of rice, represented by numerous varieties exhibiting diverse per hectare yields and physico-chemical features. Research and developmental efforts have focused on the best utilisation of paddy varieties grown in the country through:

— minimising qualitative and quantitative losses during harvest and post-harvest handling and drying stages;
— improving the milling yields of rice by appropriate drying and milling procedures, and the development of milling equipment;
— adapting processing procedures to improve nutrition and acceptability;
— utilising by-products; and
— product development for diverse needs.

The investigations at the Central Food Technological Research Institute (CFTRI) have conclusively shown that inherent grain structure and harvest/post-harvest drying practices directly influence the milling behaviour of paddy. Paddy is currently harvested at 16 to 18 per cent moisture level because of difficulties in getting labour, drying space, and other amenities during harvest periods.

Harvesting at this stage results in shattering and sun-checking of grains ultimately reflected in heavy breakages and reduced milling yields. Harvesting at higher grain moisture levels of say 20 to 24 per cent (indicated by the presence of 1 per cent milky grains), and a controlled drying procedure with an intermediate conditioning stage, is suitable for farm level use, and also reduces milling breakages. Breeding of crack-resistant varieties, the second approach, overcomes frequent process alterations and equipment designs.

Crack-resistant and low-shattering selections from pushpa, vani. and madhu varieties have been identified, tested and released by the CFTRI for mini-test trials through the University of Agricultural Sciences, Bangalore, Further, a chalky variant of Alur sanna (a rice variety), which seems to cook into discrete, fluffy grains even without ageing, has also been identified by the CFTRI for further trials. The influence of physico-chemical characteristics on the storage, packing, and processing of rice has been examined as an aid in post-harvest conservation.

For example, grains with low length: breadth ratio exhibit higher bulk density (and lower porosity), indicating that a given volume of round grains has a greater weight than slender grains. Similarly, the friction coefficient of grains increases with moisture level, contributing to handling and packing problems. Brown rice and highly milled rices, characterised by low friction coefficients, pack well, while intermediate milled rice, particularly from parboiled paddy, packs badly because of high friction and low bulk density.

Yellow discoloration of rice pre-harvest has also been a problem in some parts of the country. Varieties of rice with a high amylose content cook into dry and flaky products, while those with a low content into pasty products. The gelatinisation temperature (GT) of starch has proved to be an important quality criterion and low-GT rices are preferred for puffed products. Data such as these are extremely useful in developing varieties capable of yielding cooked products that suit the diverse food habits of Indian populations. Breeding of varieties with a higher bran oil content, and weak or no bran-lipase activity, is also indicated in relation to the edible oil requirements of the country. Breeding for low husk content would lead to higher yields of edible material, but the storage qualities of such varieties would need scrutiny.

Coarse grains consisting of jowar, maize, bajra, and other millets account for about 25 per cent of total cereal production. They are dryland crops capable of withstanding variable climatic conditions. Recent years have witnessed the evolution and commercial cultivation of hybrids, new varieties, and composites characterised by good yields, disease resistance, and adaptability to different cropping systems.

Coarse grains yield cooked products of harder texture than rice and wheat because of a fibrous bran layer that is relatively resistant to water permeability during the cooking process, and also of hard and horny subaleurone layers. Milling procedures and equipment have been evolved that remove 10 to 15 per cent of grain layers as bran to obtain grains and flours without serious impairment of nutritional qualities. Large-sized and damaged starch granules present in the grains of certain varieties of maize, jowar, and bajra are conductive to higher water uptake by flours and soft cooked products.

In making roti (unleavened bread) from coarse grain flour, the rolling of dough into circular sheets is difficult because of the absence of gluten. The steaming of dough minimises rolling problems, but detailed studies on the histo-chemistry of starch and its hydration features may help in identifying grains with desirable sheeting properties.

Over 90 per cent of the wheat produced in the country is ground in chakkis to obtain flour of about 95 per cent extraction, used for making various unleavened breads such as chapati and poor. Only about 2 million tonnes pass through roller flour mills to yield maida and white flour used in bakery products. New varieties in various stages of development and cultivation are systematically screened for characteristics needed for the manufacture of bread and biscuits.

Comprehensive data have become available from the Indian Agricultural Research Institute (IARI), the CFTRI, and the Food Grains Research Centre at Hapur. A programme has been undertaken at the CFTRI to gain an insight into the protein characteristics of wheat and their relationship to the quality of chapati, the most important means of utilising wheat in this country. Some attempts have also been made to define chapati quality by sensory methods and to correlate the data with instrumental methods of analysis. Translation of these criteria into physico-chemical characteristics could help in evolving varieties specially suited for chapati.

Pulses are the principal means of raising the protein quality of Indian cereal-based dietaries. Pulse production has not exceeded 12.5 million tonnes, although consumption requirements for 1980181 can be estimated at 15 million tonnes (at 60 9 per capita per day for a 684 million population). To minimise the wastages in traditional milling practices and hence nutrient losses, improved procedures and milling equipment have been evolved for commercial use. This technology has been used by industry for the milling of pigeon-pea almost throughout the year, independent of the climatic conditions, which play an important role in traditional practices.

Milling conditions are also being optimised for mung bean, chick-pea, urd bean, cow-pea, kidney bean, horse gram, winged bean, and peas by suitable adaptation or modification of the technology developed for pigeon-pea at the CFTRI. Wide variations in physical properties and chemical composition of each of these pulses and their varieties are to be expected in different regions and cropping systems. These properties would influence the milling behaviour of pulses, necessitating expensive changes in technologies and equipment. Intra-and inter-varietal differences in physico-chemical properties should be exploited to upgrade the milling quality of the grain, without sacrifice of yield and duration.

Storage: Food grains in the hot and humid countries of Asia suffer qualitative and quantitative losses from insects, micro-organisms and rodents during post-harvest handling and storage. Impairment of organoleptic and nutritional qualities, and health hazards arising from insect and fungal metabolises, are well-known effects of inadequate grain protection measures and undesirable storage conditions.

Food grains handled by government agencies benefit from improved storage practices, but not those remaining in farm storage. To eliminate or minimise such losses, pesticidal chemicals and their formulations are used widely for prophylaxis or destruction of insect pests both pre- and post-harvest. Pesticide residues are monitored and regulated under Indian food laws to ensure safety to consumers.

Increasing hazards from the pesticide residues on food grains accumulating in the human system. either by direct consumption or through animal-based foods, have led to an intensification of research on non-toxic insecticides, biological methods of control, and breeding of varieties resistant to pre-harvest infestation. A mass of experimental data on the varietal resistance of different food grains to storage pests has been put out from various parts of the world.

More often than not, the susceptibility or resistance of high-yielding varieties to storage pests is tested just before release for commercial cultivation. Attention to pest resistance is important at all stages of the breeding programme. Maintaining desirable genetic characteristics on a national scale of cultivation is difficult and expensive. Sixty-seven varieties of rice have been tested for field infestation by Angoumois paddy moth (Sitotroga cerealella) and graded for susceptibility.

Information on susceptibility or resistance of wheats grown in India to storage pests such as the rice weevil (Sitophilus oryzae), or khapra beetle (Trogoderma granarium), Rhizopertha dominica, and Tribolium castaneum have also been generated to help in breeding and cultivation. In maize, Angoumois paddy moth is found to damage varieties with a high amylose content.

Reports from Africa indicate that local varieties are more resistant to insect infestation than improved varieties and hybrids, because of hard kernels and complete coverage of cobs by sheaths. Resistance and susceptibility of India maize varieties to Trogoderma granarium and Rhizopertha dominica have also been tested. Six genotypes (M-25-1, CSH-1, CSH-2, CSH3, CSH-4, and CSH-5) of Indian jowar have also been tested and the degrees of resistance to storage pests ascertained.

Varietal resistance of pulses to storage pests has also been indicated in India. Trials on chick-pea have shown that grains of indigenous varieties G-24 and G-30 are more tolerant than new varieties to the pulse beetle (Callosobruchus chinensis) because of a wrinkled surface and tough coat. Continuous efforts are necessary to develop and maintain pest-resistant germplasm of different food grains so as to economise on the costs of storage and infestation control.

Breeding for Nutritional Quality

Breeding for nutritional quality has mainly been directed to improving the protein content, which may or may not be reflected in increased levels of specific essential amino acids. The dwarf wheats cultivated in India contain 12.5 - 16 per cent protein and 2.21 - 2.9 9 of lysine per 100 9 protein. The protein content of improved sorghum varies from 7 to 10 per cent, and that of lysine from 0.9 to 2.6 9 per 100 9 protein. Lysine is the amino acid that limits the nutritional quality of sorghum protein; however, there is overwhelming evidence to show that excess of dietary leucine (another essential amino acid) results in a conditioned nicotinic acid deficiency.

More recent studies have shown that the leucine/isoleucine ratio may be as important as the leucine level in causing pellagra. Deficiency of lysine is also relevant in bajra and ragi as in other cereals. Tannin present in the testa of sorghum grains at about 1 per cent level is known to reduce protein digestibility and availability by binding some of its fractions. Most Indian varieties contain less than 0.1 per cent tannin and it may not be of much significance. However, recently the National Institute of Nutrition indicated that the big-availability of iron from a variety of jowar that contained 136 mg/100 9 of tannin was lower than in one containing 20 mg/100 g of tannin.

Of the common cereals, the big-availability of iron from jowar is least and iron deficiency anaemia is common among poorer sections of the population. A low tannin content in jowar is therefore a desirable breeding feature. Very little attention seems to have been given to the carbohydrates of food grains be they cereals, millets, or pulses, although animal experiments indicate that flatulence can occur by feeding food grains containing stachyose, and that protein quality can be affected by the level of digestibility of dietary starches. For example, feeding the field bean and some of its carbohydrate fractions to rats affected the absorption of calcium and nitrogen. Detailed carbohydrate profiles of food-grain varieties appear relevant in breeding programmes in terms of nutrition, processing and culinary practice.

The spectre of hunger, poverty, and malnutrition continues to stare at mankind. It represents a multidisciplinary challenge of no small magnitude and therefore requires a multidisciplinary approach to find a solution. Science and technology have been able to make meaningful contributions to socioeconomic development only when they have acted in an interdisciplinary manner to solve the problems.

The United Nations University has therefore recognised the value of such an approach and has given special attention to organising activities that would involve teams of scientists (both social and natural), technologists, policy-makers and planners (including development economists) and the implementers of programmes to collectively look into the major problems of mankind and find solutions for them through co-operative efforts.

The United Nations University is doing this in the hope that the concerned disciplines will stimulate each other consciously and create a comprehensive and dynamic system capable of multidisciplinary action that could increase the pace of progress towards establishment of a more equitable and just social order in this world. This effort could convert the vicious circles in which we are caught at present into dynamic development cycles.

UN University has supported the organisation of interface workshops. It is hoped that the present workshop on the interfaces between agriculture, nutrition, and food science will provide an opportunity for better understanding of the whole system in greater depth and in relation to actual problems, so that more meaningful multidisciplinary solutions can be sought through co-operation between the scientific communities concerned in other fields.

The world population, which at present stands at about 4,300 million, is expected to reach a figure of well over 6,000 million by the year 2000. Agricultural production has barely kept pace even with the present need. The requirement of food, even to meet minimum need, will be nearly twice the present production level by the turn of the century and the challenges for the next century would become much greater. Food losses continue to be high and take away from mankind a substantial amount of what is produced with a great deal of inputs and human effort. The consequent qualitative deterioration of food, resulting from infestation by rodents, insects, and micro-organisms, adds to the problem of malnutrition.

Prevention of these losses would increase and improve food supplies without additional demand on land, and raise nutritional standards. The UN University looks to this workshop to provide multidisciplinary leadership, and for its recommendations that may be useful in moving forward more rapidly towards solving this global problem, which is among the greatest challenges facing mankind in the twenty-first century, scientifically, socially, and politically. In this, the efforts of the organs, organisations and bodies of the United Nations system will have an important role to play, but the real efforts needed will have to come from the countries themselves where the problem really exists.

Food-Energy Systems

The relation between food and energy problems first became evident as a result of the oil crisis in the early 1970s. While immediate attention was given by industrialised countries to ensuring adequate oil supplies to fuel their energy-intensive food systems, long-term concerns were raised about the plight of the rural and urban poor in third world countries with the realisation that the high cost of energy and fertilisers would further limit the scope of the Green Revolution.

Beyond the oil price problem loomed the second energy crisis, with even greater social and ecological consequences for more than half of the world's population. In practically all third world countries the problems of getting food to eat began to be overs had owed by the problems of acquiring the energy needed to cook it. Apart from the financial sacrifices, there was a severe strain on time budgets, notably those of women and children, who spend increasingly long hours collecting fuelwood.

These problems are exacerbated by the seasonal imbalance in biomass supply and the vicious cycle of greater quantities of dung being used as fuel rather than as fertiliser for maintaining crop production. Rising fuel prices, boosting transportation and agricultural costs, will inevitably push food prices beyond the reach of hundreds of millions of already hungry people. Rising populations, despite the best efforts to reduce fertility rates, will continue to increase the demand for both food and energy. The developing countries will not be able to solve their food problem without solving their energy problem and, without a satisfactory solution to both their economic growth will be severely constrained. The centrality of the food and energy nexus calls for a comprehensive policy approach.

Integrated food-energy systems as a catalyst for rural development and industrialisation; and alternative urban development strategies based on greater self-reliance. FEN is predicated on the idea that positive synergies can be developed by addressing simultaneously the ideas of production and access to food and fuel and by building around these twin objectives self-reliant development strategies. The broad scope of FEN and the modest resources that it had available pointed naturally to the significance of its "enzyme" role.

Special attention was paid to the "third system" or citizens' organisations through, inter alia, the urban self-reliance project that was implemented with the International Foundation for Development Alternatives (IFDA) in Nyon, Switzerland. From the very beginning, an effort was made to give a global learning dimension to FEN and to emphasise South-South co-operation. This began with a study tour by four Brazilian researchers to Senegal, India, and China and was facilitated by FEN conferences that were subsequently held in Latin America, Asia, and Africa. This policy of strengthening research capacities in third world countries and improving communication between people working on similar issues was also adopted by FEN projects in the urban field.

Food and energy are global problems in at least four ways. First, their regular and continuing availability is a condition sine qua non of human survival, posing a formidable challenge that must be tackled simultaneously from both the supply and demand sides. Food and fuel stocks are of no help and little consolation to people who cannot afford to buy them and have no access to the resources needed to produce them.

Second, assuming optimistically that humanity will manage to solve the problems posed by its bare survival, the quality of life of millions of people will still depend to a great extent on increased supplies and better use of both food and energy; their central role in a need-oriented development strategy is only too obvious.

Third, both food and energy loom large in the North-South confrontation as potential weapons and tools of domination; hence the importance of global negotiations to modify the present gloomy picture and bring about some constructive international co-operation in both fields.

Finally, food and energy production affect and are affected by the state of the environment: energy-, land-, and water-use patterns will increasingly influence the climate and other aspects of our life-support systems. This will have far-reaching consequences for the long-term prospects of food and energy production in semi-arid areas and threatens the very existence of flood-prone coastal settlements, home to millions of people.

Integrated Food-energy Systems

One of the two major focuses of FEN was the analysis of systems designed to integrate, intensify, and thus increase the production of food and energy by transforming the by-products of one system into the feedstocks for the other. Such integrated food-energy systems (IFES) can operate at various scales, ranging from the industrial-sized operations in Brazil designed to produce primarily ethanol and fertilizer, to the village-and even household-level biogas systems in India.

A conceptual outline of such systems, including the long-ignored but vitally important component of water for biomass-based production. This valorisation of agricultural by-products plays an important role in the design of IFES for specific agro-climatic regions, designs which should closely follow the paradigm of natural ecosystems. These modern, ecologically sound systems are characterised by their closed loops of resource flows.

In a sense, they incorporate the same rationality of traditional peasant farmers, albeit at a completely different level of scientific and technical knowledge. The word "integrated" has three superimposed meanings: multiple sources of inputs (including energy), multi-task devices, and time sharing of devices (such as engines).

Techniques of different vintages are also used in an attempt to accommodate heterogeneity rather than to impose a homogenising modernity across the board resulting in a massive displacement of labour. The design of IFES requires a simultaneous consideration of the biophysical components of resource management, of the social and ecological impacts of the technologies used, and of the institutional settings involved.

For each site-specific configuration of climatic and environmental conditions, several socially desirable, ecologically sustainable, and economically efficient production systems are conceivable, differing in output mix, forms of social organisation and community participation, size of operation, complexity of design, and technical sophistication. Ideally, they should have a modular structure, allowing for a progressive implementation by adding new modules to the initial structure.

By comparing several systems, observing their performance, and exchanging experiences and results between projects situated in similar ecosystems and different cultural areas, development planners would come as near to a social laboratory setting as it is possible. In order to adapt proposed solutions to the specific local conditions and needs, community involvement is required at all stages: to identify not only the pressing problems but also the latent resources, then to progressively build the system, and finally to manage it in a proactive way.

Special attention should also be given to integrated systems designed for ecologically vulnerable areas and for reclaiming wastelands. Socially responsive and ecologically sustainable agroforestry systems, protecting both the local people and the trees, should receive maximum attention.

As for wastelands, considerable scope exists for establishing fish farms as part of mixed farming systems. In many maritime states of India, for example, large areas of saline soils, marshy swamps, and mangroves are lying fallow. Their reclamation for agriculture may be costly but such resources could be used for brackish water aquaculture generating both food and work for landless peasants.

IFES as the adoption of agricultural and industrial technologies that allow maximum utilisation of by-products, diversification of raw materials, production on a small-scale, recycling and economic utilisation of residues, and harmonisation of energy and food production. Such systems imply the need for comprehensive land-use planning as well as planned interrelations among soil, water, and forest resources in relation to agricultural residues. Major advantages are their minimal negative environmental impact, and their decentralisation and efficiency, which often have positive social and economic side-effects.

In Brazil, IFES notably include micro-distilleries for alcohol production from sugar cane, sweet sorghum, maniac or sugar beets; fattening of stall-fed livestock; biodigestors for decomposition of livestock manure and/or sugar cane bagasse and/or stillage; generation of electricity and agricultural mechanisation on the basis of the fuels thus produced; and the application of biofertiliser to cultivated lands. The possibility of extending this to include various ways of enhancing the water management system, including the production of aquatic plants, fish, and zooplankton as well as the retention of water for sedimentation and irrigation purposes.

IFES can be put together in various configurations and at various scales. A fundamental distinction, however, can be made according to their ultimate purpose. One kind is "farm-centred", or in the case of agribusiness, enterprise-centred, although any surpluses might be used to satisfy the non-production needs of workers or farmers. This is not the purpose of the system, however; it is only a spin-off.

Another system is the "energy farm" unit designed for the production of energy, usually for distribution via conventional means to distant urban markets. This type of system could be expanded into a kind of "public utility" system in order to include a social purpose other than food production, for example, waste water treatment in a manner that simultaneously produces food and reduces the environmental load. An urban latrine system in India that, coupled with a biogas generator, produces both hot water and street lighting while reducing the sewage treatment problem.

A third type of IFES is the "community focused" system. It seeks to energize daily life in a variety of ways that answer domestic and community needs, such as cooking and sanitation, as well as individual and community productive needs in agriculture and industry. The features of this type of system as including:

— using a technology mix designed as a minimum cost alternative;
— meeting energy needs not only for agriculture, but also for other social needs;
— maximising utilisation of available big-resources with minimum harm to the environment;
— benefiting all classes of the community;
— increasing food productivity;
— generating additional employment and off-farm income; and
— requiring minimal maintenance so that villagers themselves can operate and manage it.

IFES do not necessarily offer a fundamental departure from conventional models of "trickle down" and "modernisation" development. Thus those designers who wish to put big-energy systems at the service of "another development" should pay close attention to the local patterns of food and energy provisioning in addition to other basic needs such as clean water.

The same situation applies to the ability of IFES to balance economic, social, and ecological evaluations. Although an analysis shows that environmental degradation can be prevented by using IFES, it remains to be seen whether or not this is actually happening. This raises the question of whether or not IFES that are environmentally sustainable can be promoted instead of energy-intensive agricultural modernisation projects that aggravate social and economic conditions.

To date, most IFES research that has been done suffers from an engineering bias. Although this provides systems with a solid foundation, notably from the biological and social sciences. While most systems take advantage of biofertilisers, there seems to be a "black box" approach to the material that goes into and comes out of biodigestors.

The Chinese Energy Village described by Zhang et al. is a community-oriented system that seems to transcend some of these problems. Because it was designed in co-operation with the residents of a "natural village", the results allow for a complex, transitional pattern of energy use. Coal, diesel, and kerosene continue to share a place with the new and old renewable energy sources.

Production techniques for both market and subsistence needs are addressed simultaneously by this system. The latter include both family requirements (lighting, cooking, space heating) and social purposes (new fuels). Both productive and non-productive uses are thus served and there seem to be advantages for everyone in the village. Unfortunately, such a balanced system is hard to obtain in real life as IFES are not sealed off or protected from the rest of the economy (and this is increasingly the case even in China).

Among the early initiatives of FEN was an analysis of the entire chain of food production activities "from the farm gate to the food plate". It naturally paid special attention to the intersections with energy flows, the by-products produced and the losses at each stage, thereby focusing attention on ways to increase efficiency by reducing wastes. There are five basic steps:

— description of the system to be studied;
— an energy analysis of the system;
— identification of possible technological paths for the supply of energy to it;
— an assessment of alternative technological paths; and
— a plan for the introduction of the chosen project.

Given the complexity of such an analysis, there is usually a trade-off between the quantity of detailed information one would like to have and the amount one can afford, given constraints on money, personnel, and time. One must thus determine the minimum amount of information necessary to understand an existing agro-energy system well enough to make an intelligent intervention.

Among the social impacts of IFES, attention should be given to highly seasonal employment patterns, particularly in large-scale sugar cane production, and the poor quality of life of rural families living on vast plantations. The inclusion of other crops and economic activities in these systems would help to provide year-round employment. A related concern is the consequence of "creating" a wide variety of "new resources" in rural areas. More emphasis also needs to be given to the impacts of IFES on women and children.

Attention should be given here to the potential of IFES to contribute not only to the sustainability of rural resource-use patterns, but to their "regeneration".

The word 'sustainable', which is frequently used in reference to new agricultural futures, too often is interpreted to mean that, given necessary resources, even a poor system can be sustained for a long time, provided only that a community has the ability to obtain the needed resources. To move beyond this ambiguity, the word regenerative is used.

The idea of regenerativeness goes beyond conceptualisations of conservation, for this latter word usually just conjures up the idea of being careful about using a resource in order to extend its time horizon as much as possible. Regeneration, in contrast, and particularly in the case of agriculture, refers not only to the replacement of the essential resource, but, hopefully, to its enhancement.

The natural resource base by restoring stability and integrity in both the socio-economic and biophysical sense. The possibility of applying this approach to other regions suffering from desertification and/or conflicts between food and energy production. The dehesa is the result of centuries of interaction between farmers and their environment which has resulted in open, savanna-like woodlands with a rich diversity of flora and fauna, thus giving it stability and resilience.

It allows sustainable harvesting of fuelwood from periodical pruning and clearing, which also improves the quality of the trees and pastures, thus increasing food production. Apart from cereal crops for human consumption, dehesas mostly produce fodder crops to supplement the natural grazing resources. The carefully cultivated oak trees provide not only fuelwood but forage and acorns for livestock fattening in addition to cork, which is used as an industrial feedstock. An agro-silvo-pastoral combination producing food, energy and forest products is possible and can provide a balanced diet and all the products rural populations may need.

This resource-use pattern shows that sustainable resource exploitation is not only compatible with ecosystem regeneration but actually part and parcel of it. The different environmental and socio-economic characteristics of each region necessitate specific solutions and it must be borne in mind that the dehesa example cannot be transported per se to other areas without taking into account these considerations.

Agricultural Refineries

The situation of third world countries can perhaps best be understood by an examination of the experience, both positive and negative, accumulated in this respect by Asian and Latin American countries. The two largest, land-scarce countries, China and India, are still predominantly rural, in contrast with Brazil, where the abundance of land goes hand in hand with premature and very costly urbanisation. Both China and India have an excess labour force in agriculture; their problem, therefore, is to promote the agricultural exodus.

The official Chinese policy is aimed at releasing some 150 million workers from agriculture by the end of the century. The slogan is thus: "The Chinese peasant leaves agriculture but stays in the village". As for India, Chakravarty makes it clear that in spite of the Green Revolution and surplus food production, agriculture cannot absorb fully the surplus labour even if the existing regional unbalances in agricultural growth are overcome. On the other hand, neither can industry absorb the migrant labour force even with a high rate of industrialisation.

It would, therefore, be necessary to adopt policies which would generate adequate off-farm employment in the rural areas and small towns so that migration to urban and metropolitan cities and the consequent accentuation of urban social problems could be avoided. The basic problem, therefore, for all developing countries with a rapidly growing population and increasing rural-urban migration, both permanent and seasonal, is how to promote off-farm migration through "industrialisation without depeasantisation".

For China and India, this is a necessity, and for a country like Brazil, it is an opportunity to use land productively while avoiding the unnecessary costs of excessive urbanisation. Rural industrialisation is usually associated with intermediate or even simpler technologies. But it need not and should not necessarily be so. Recent advances in communication, agricultural, and chemical technologies have made the problem of economies of concentration and scale somewhat obsolete because they allow many secondary and tertiary activities to be deployed in rural areas.

To the extent to which underdevelopment is a "co-existence of asynchronisms", selective modernisation, based on an endogenous project, calls for skill-ful management of technological pluralism. No developing country can sustain across the board the deadly rhythm of accelerated obsolescence imposed by competition on world markets. Neither can it afford technological stagnation. Hence the need to plan sectoral modernisation and to design public policies that can offset the homogenising effect of market forces.

The relevance for rural areas of decentralised IFES based on biomass should be understood. Such industrialisation should start with using existing resources and raw materials. Agricultural residues and biomass production are the obvious choices. In tropical countries, this is particularly true for "agricultural refineries" designed to produce a broad range of industrial products from biomass. The key element in this alternative rural development strategy is a process based on valorizing biomass as a dynamic anchor activity around which organisations and other development activities can work.

The "agricultural refinery" concept can be illustrated by the case of sugar cane production. Its central anchor activity is a modern processing industry around which all other activities could be oriented. An average sugar factory with a 2,500 tonne-per-day capacity in India commands an area of about 10,000 to 12,000 ha spread over 30 to 40 villages, employing about 700 to 800 people on a permanent basis and about 2,500 people on a seasonal basis for about 150 to 180 days per year. It also generates secondary and tertiary employment in the service sector. The full potential of such an industry, however, can only be achieved if all of the possibilities of by-product processing are exploited.

This alternative is particularly relevant to Brazil where its National Alcohol Programme is the largest such endeavour in the world, with over 11 billion litres produced per year. But the current low oil prices are causing intense controversy over its cost-effectiveness. The rationalisation in the use of existing distilling capacity by improving the productivity of the entire plantation is thus imperative. Turning alcohol distilleries into "agricultural refineries" for overall sugar cane processing is one of the best ways towards this end.

This is being shown by experiments now underway in some installations where sugar cane bagasse is being used as an input for cattle raising and electricity generation. It should be noted that a similar "agricultural refinery" system can be designed for almost all agricultural commodities, such as rice, wheat, cotton, maize, etc. Such IFES would obviously depend on the agricultural and other natural resources available from local sources in the area concerned.

Given recent advances in science and technology, particularly in chemical processing and biotechnology, the exploitation of biomass for IFES opens up new vistas for alternative development. in fact, with the emergence of such a rural industrialisation process, and new rural-urban configurations, one can visualise the end of the costly urbanisation process and a chance for third world countries to leap-frog into the 21 st century.

Food-Energy Nexus

The emergence of "sideline", "real" or "parallel" markets and activities de mands a reassessment of how one carries out "village ecosystem" studies. Although impressive competence at estimating energy flows in villages has been developed over the past few years, there still is little formal understanding of these micro-macro connections. Energy-based definitions of "efficiency" are certainly important, but they need to be complemented with the viewpoints of local people based on their daily lives. Otherwise, there is a distorting influence that leads one away from questions of food quality, welfare, and health benefits of alternative uses of a carbohydrate source such as sugar cane.

Another methodological theme to emerge was the value of "participatory action research" that exploits local knowledge of micro-environments and emphasises the importance of building on the skills and knowledge that exist in villages. There is no doubt that everything is in place for large-scale, rapid progress in meeting human needs through IFES innovations. A biomass production/industrialisation scenario for India would generate 2,000 million work days of employment and 90 million tonnes of charcoal, serve as a reliable energy base for rural industrialisation, and put 20 million ha of wasteland under forest cover, while still being a strictly bankable proposition.

Other elements for rapid progress are also available. the engineering side of IFES has clearly reached a degree of maturity with designs, prototypes, and functioning systems now available. Two missing links, needed to make effective use of this experience, have begun to be forged: rapid advances in biotechnology and breakthroughs in the social organisation of effective community participation. What is needed is more focus by IFES proponents on integrating their work with these two links.

What is also needed is to translate the political will at a governmental level into the first steps in actually reworking some of the distorting macro-economic and social policies inherited from the past. Such action need not be all that revolutionary but could include some steps towards reforming price, taxation, and subsidy policies needed to implement IFES systems. Another important move would be to give concrete meaning to the verbal commitment to women's equality through such things as day care facilities for rural women engaging in new employment schemes.

The second major focus of FEN was on ways to increase access to food and energy by the urban poor through encouraging innovative, resource-conserving, and employment-generating urban development strategies. FEN's urban projects shared many points in common with the "resource-conserving" cities of Meier and the Managing Energy and Resource Efficient Cities (MEREC) programme of USAID.

Urbanisation is by far the most important social transformation of our times. In 1800 no more than 3 per cent of the world's population lived in cities, but by the year 2000 urban dwellers will outnumber the rural population. Furthermore, most of the increase will occur in third world cities. The urban population in third world countries will have multiplied almost 16 times, from less than 200 million to 3,150 million people. By comparison, the urban population in industrialised countries multiplied only about five times from 1840 to 1914, the period of its most intensive growth.

The consequences of these trends are assessed in diametrically opposed fashion by the supporters and the foes of large cities. The former emphasise the civilising role of cities, the high productivity achieved by industries and modern services thanks to their unprecedented degree of concentration, the amenities of urban areas (in sharp contrast to the apparent drudgery of rural areas and smaller centres), and the multiple opportunities for work and self-realisation offered to their inhabitants.

The latter insist on the parasitic character of the city, diverting and draining for its own advantage the economic surplus produced by the countryside. They point to the deep disruption of the urban environment with the attendant health hazards, the often appalling housing and working conditions of the urban poor, the endemic unemployment and underemployment, and the social anonymity resulting from sub-human living conditions.

The wonders achieved by cities with the economic surplus that they have been able to concentrate, one should not forget that primitive accumulation was largely through extracting this surplus from the peasantry of today's industrialised countries and their colonies.

At stake for third world countries is the opportunity to transform their condition of "lateness" into an advantage. Modern science and technology associated with a critical analysis of the impasses of industrialised societies should allow third world countries to find alternative patterns of urbanisation and to implement them at far less social, economic, and ecological cost.

This is not to say that substantial progress could not be achieved in third world countries through more traditional, Western-style methods. But the magnitude of the financial effort required makes this proposition unrealistic in the present political context, even if the volume of necessary investment does not exceed the theoretical possibilities of the world economy.

Yet it may also be said that urbanisation cannot be carried out along the old lines, following the model of the large cities of the industrialised countries. To put it more precisely, in third world countries the pattern of city growth reflecting that of the industrialised countries—and the resultant increase in the speed of urbanisation, render it practically impossible to cater to the basic requirements of the majority of third world residents.

To simply house the additions to the urban population between now and the end of the century, the equivalent of over 600 cities of one million residents each would need to be built. An exhaustive study of the metabolism of Hong Kong by UNESCO's Man and the Biosphere programme (MAB) concluded that the energy costs of constructing and operating such cities between 1978 and the year 2000 would be five times higher than total world consumption of energy in 1973.

Finally, one ought to mention the lack of appropriate policies to implement such an ambitious construction programme and to apply resource conservation principles in an equitable manner. Urban studies a certain degree of consensus exists about macro-problems, but nothing of the kind happens with respect to micro-problems which affect low income groups more directly. There is a visible gap between the effort made to arrive at the diagnosis and the neglect in creating the necessary policy instruments.

The ways in which market and non-market economic activities combine are quite complex. They constitute the fabric of the "real" economy. Non-market economic activities should not be viewed as a residual category designed to disappear with technical progress. On the contrary, their share in terms of time allocation may increase, pari passu, with the reduction of the working time against wages, as a matter of deliberate choice. The market/non-market dichotomy therefore offers a useful starting point to move in the direction of a development theory not exclusively based on the categories of the market economy and on monetary metrics.

The informal sector is often referred to as "hidden" or "invisible". A better term would be "statistically unrecorded" as most of the activities in the market segments encompassed by these names are conducted out in the open. Neither the "formal/informal" nor the "open/hidden" dichotomy offers a suitable framework to describe the latticework of the real economy. Moreover, they lend themselves to statistical manipulation, as both the informal and hidden sectors are in reality residual categories.

By definition, they include everything that has not been specifically recorded as belonging to the narrowly defined organised sector of the commoditised economy. By postulating that all those who do not find a regular job in the organised sector are absorbed by the informal economy, it becomes possible to assume away the problems of unemployment and underemployment, as well as to play down the disruptive social consequences of the emergence in industrialised countries of a "two-speed" economy: a highly performing and competitive sector open to a minority, and a residual one for the rest of the population.

The large cities thus prove to be cities for the elite which, at best, function as mechanisms for the regressive redistribution of investments and wealth. The poor end up by paying more dearly than the rich for the basic services, without which life would be impossible. The extreme case is illustrated by cities like Karachi where the inhabitants of poor neighbourhoods have been known to pay itinerant water-bearers 20 times more for their (questionable) water supply than is paid by the inhabitants of the rich neighbourhoods where running water is supplied.

In contrast to the commoditised economy, household activities are situated both outside the labour and product markets, although household consumption obviously consists of a bundle of goods and services that are both purchased on the market and self-produced.

In addition, the non-market segment of the economy also comprises the social sector, consisting of all collective activities organised outside the market by neighbourhood and community groups, citizen associations, and, in some cases, co-operatives. Their common trait is that they are founded, just like the household sector, on the principle of reciprocity: the donation of an unpaid productive activity, deemed of social interest and matched by free competition of goods and services collectively produced.

A closer look at the real economy of the city (including the multiplicity of interconnected markets, ranging from the legal to the criminal, non-market household activities, and state intervention through to subsidies, rationing, distribution of goods, and provision of services) shows that many of these resources, not accounted for in official statistics, are being intensively used by people to build their homes, produce some food for self-consumption or sale, and transform recycled waste into saleable commodities.

For the bureaucrats from the ministries of planning or finance, the answer is clear: austerity budgets compounded by the requirements of foreign debt servicing mean that the so-called "nonproductive" investments and collective consumption must be cut to the bone. Whatever the outcome of the debate between the partisans and critics of International Monetary Fund-style policies, one thing is clear. Difficult as it is, the situation is not entirely stalemated so long as there are idle, underutilised, or dilapidated physical and human resources that can be used to produce socially desirable goods and services without violating the prevailing budgetary restrictions.

A bootstrap operation would be based on reducing waste in order to increase the resources available for development. It would be aimed at improving living conditions in poor urban areas through grassroots community action without waiting for massive funding from outside. Such an operation has, of course, obvious limitations. It cannot by itself solve the economic crisis and generate enough jobs to reabsorb the backlog of unemployed. Furthermore, under no circumstances should it be used as an excuse for the authorities at local, regional, and national levels to shirk their responsibilities in this field.

On the contrary, grassroots action must be actively supported. Self-reliance does not necessarily mean self-sufficiency any more than an inward-looking development strategy leads to delinking. The complexity of the modern world cannot be tackled by decomposing it into an archipelago of self-sufficient communities, be they rural, urban, or rurban. But self-reliance in moral, political, and intellectual terms makes people resourceful and confident: they assume their situation instead of taking a passive approach; looking around them, they end up by identifying resources in their own backyard that can be exploited to bring some relief to their plight.

References

Curley, S., and Mark (1990). *The Natural Guide to Good Health*, Lafayette, Louisiana, Supreme Publishing

Galdston, I. (1960). *Human Nutrition Historic and Scientific.* New York: International Universities Press.

Mahan, L.K. and Escott-Stump, S. eds. (2000). *Krause's Food, Nutrition, and Diet Therapy* (10th ed.). Philadelphia: W.B. Saunders Harcourt Brace.

Thiollet, J.-P. (2001). *Vitamines & minéraux.* Paris: Anagramme.

Walter C. Willett and Meir J. Stampfer (January 2003). "Rebuilding the Food Pyramid". *Scientific American* 288 (1): 64–71.

2

Basic Human Nutrients

The diet of an organism is what it eats, which is largely determined by the perceived palatability of foods. Dietitians are health professionals who specialize in human nutrition, meal planning, economics, and preparation. They are trained to provide safe, evidence-based dietary advice and management to individuals (in health and disease), as well as to institutions. Clinical nutritionists are health professionals who focus more specifically on the role of nutrition in chronic disease, including possible prevention or remediation by addressing nutritional deficiencies before resorting to drugs. While government regulation of the use of this professional title is less universal than for "dietician", the field is supported by many high-level academic programs, up to and including the Doctoral level, and has its own voluntary certification board, professional associations, and peer-reviewed journals, e.g. the American Society for Nutrition, Nutrition Society of India, Food Scientists and Nutritionists Association India, Indian Dietetic Association and the American Journal of Clinical Nutrition.

A poor diet may have an injurious impact on health, causing deficiency diseases such as scurvy and kwashiorkor; health-threatening conditions like obesity and metabolic syndrome; and such common chronic systemic diseases as cardiovascular disease, diabetes, and osteoporosis.

Nutritional science investigates the metabolic and physiological responses of the body to diet. With advances in the fields of molecular biology, biochemistry, nutritional immunology, molecular medicine and genetics, the study of nutrition is increasingly concerned with metabolism and metabolic pathways: the sequences of biochemical steps through which substances in living things change from one form to another.

Carnivore and herbivore diets are contrasting, with basic nitrogen and carbon proportions being at varying levels in particular foods. Carnivores consume more nitrogen than carbon while herbivores consume less nitrogen than carbon, when an equal quantity is measured.

The human body contains chemical compounds, such as water, carbohydrates (sugar, starch, and fiber), amino acids (in proteins), fatty acids (in lipids), and nucleic acids (DNA and RNA). These compounds in turn consist of elements such as carbon, hydrogen, oxygen, nitrogen, phosphorus, calcium, iron, zinc, magnesium, manganese, and so on. All of these chemical compounds and elements occur in various forms and combinations (e.g. hormones, vitamins, phospholipids, hydroxyapatite), both in the human body and in the plant and animal organisms that humans eat.

The human body consists of elements and compounds ingested, digested, absorbed, and circulated through the bloodstream to feed the cells of the body. Except in the unborn fetus, the digestive system is the first system involved. In a typical adult, about seven liters of digestive juices enter the lumen of the digestive tract. These digestive juices break chemical bonds in ingested molecules, and modify their conformations and energy states. Though some molecules are absorbed into the bloodstream unchanged, digestive processes release them from the matrix of foods. Unabsorbed matter, along with some waste products of metabolism, is eliminated from the body in the feces.

Studies of nutritional status must take into account the state of the body before and after experiments, as well as the chemical composition of the whole diet and of all material excreted and eliminated from the body (in urine and feces). Comparing the food to the waste can help determine the specific compounds and elements absorbed and metabolized in the body. The effects of nutrients may only be discernible over an extended period, during which all food and waste must be analyzed. The number of variables involved in such experiments is high, making nutritional studies time-consuming and expensive, which explains why the science of human nutrition is still slowly evolving.

In general, eating a wide variety of fresh, whole (unprocessed), foods has proven favorable for one's health compared to monotonous diets based on processed foods. In particular, the consumption of whole-plant foods slows digestion and allows better absorption, and a more favorable balance of essential nutrients per Calorie, resulting in better management of cell growth, maintenance, and mitosis (cell division), as well as better regulation of appetite and blood sugar. Regularly scheduled meals (every few hours) have also proven more wholesome than infrequent or haphazard ones, although a recent study has also linked more frequent meals with a higher risk of colon cancer in men.

Classes of Nutrients

There are six major classes of nutrients: carbohydrates, fats, minerals, protein, vitamins, and water.

These nutrient classes can be categorized as either macronutrients (needed in relatively large amounts) or micronutrients (needed in smaller quantities). The macronutrients include carbohydrates (including fiber), fats, protein, and water. The micronutrients are minerals and vitamins.

The macronutrients (excluding fiber and water) provide structural material (amino acids from which proteins are built, and lipids from which cell membranes and some signaling molecules are built) and energy. Some of the structural material can be used to generate energy internally, and in either case it is measured in Joules or kilocalories (often called "Calories" and written with a capital C to distinguish them from little 'c' calories). Carbohydrates and proteins provide 17 kJ approximately (4 kcal) of energy per gram, while fats provide 37 kJ (9 kcal) per gram., though the net energy from either depends on such factors as absorption and digestive effort, which vary substantially from instance to instance. Vitamins, minerals, fiber, and water do not provide energy, but are required for other reasons. A third class of dietary material, fiber (i.e., non-digestible material such as cellulose), is also required, for both mechanical and biochemical reasons, although the exact reasons remain unclear.

Molecules of carbohydrates and fats consist of carbon, hydrogen, and oxygen atoms. Carbohydrates range from simple monosaccharides (glucose, fructose, galactose) to complex polysaccharides (starch). Fats are triglycerides, made of assorted fatty acid monomers bound to a glycerol backbone. Some fatty acids, but not all, are essential in the diet: they cannot be synthesized in the body. Protein molecules contain nitrogen atoms in addition to carbon, oxygen, and hydrogen. The fundamental components of protein are nitrogen-containing amino acids, some of which are essential in the sense that humans cannot make them internally. Some of the amino acids are convertible (with the expenditure of energy) to glucose and can be used for energy production, just as ordinary glucose, in a process known as gluconeogenesis. By breaking down existing protein, some glucose can be produced internally; the remaining amino acids are discarded, primarily as urea in urine. This occurs normally only during prolonged starvation.

Other micronutrients include antioxidants and phytochemicals, which are said to influence (or protect) some body systems. Their necessity is not as well established as in the case of, for instance, vitamins.

Most foods contain a mix of some or all of the nutrient classes, together with other substances, such as toxins of various sorts. Some nutrients can be stored internally (e.g., the fat soluble vitamins), while others are required more or less continuously. Poor health can be caused by a lack of required nutrients or, in extreme cases, too much of a required nutrient. For example, both salt and water (both absolutely required) will cause illness or even death in excessive amounts.

Carbohydrates

Carbohydrates may be classified as monosaccharides, disaccharides, or polysaccharides depending on the number of monomer (sugar) units they contain. They constitute a large part of foods such as rice, noodles, bread, and other grain-based products. Monosaccharides, disaccharides, and polysaccharides contain one, two, and three or more sugar units, respectively. Polysaccharides are often referred to as complex carbohydrates because they are typically long, multiple branched chains of sugar units.

Traditionally, simple carbohydrates were believed to be absorbed quickly, and therefore to raise blood-glucose levels more rapidly than complex carbohydrates. This, however, is not accurate. Some simple carbohydrates follow different metabolic pathways which result in only a partial catabolism to glucose, while many complex carbohydrates may be digested at essentially the same rate as simple carbohydrates. Glucose stimulates the production of insulin through food entering the bloodstream, which is grasped by the beta cells in the pancreas.

Fibers

Dietary fiber is a carbohydrate (or a polysaccharide) that is incompletely absorbed in humans and in some animals. Like all carbohydrates, when it is metabolized it can produce four Calories (kilocalories) of energy per gram. However, in most circumstances it accounts for less than that because of its limited absorption and digestibility. Dietary fiber consists mainly of cellulose, a large carbohydrate polymer that is indigestible because humans do not have the required enzymes to disassemble it. There are two subcategories: soluble and insoluble fiber. Whole grains, fruits (especially plums, prunes, and figs), and vegetables are good sources of dietary fiber. There are many health benefits of a high-fiber diet. Dietary fiber helps reduce the chance of gastrointestinal problems such as constipation and diarrhea by increasing the weight and size of stool and softening it. Insoluble fiber, found in whole wheat flour, nuts and vegetables, especially stimulates peristalsis – the rhythmic muscular contractions of the intestines which move digesta along the digestive tract. Soluble fiber, found in oats, peas, beans, and many fruits, dissolves in water in the intestinal tract to produce a gel which slows the movement of food through the intestines. This may help lower blood glucose levels because it can slow the absorption of sugar. Additionally, fiber, perhaps especially that from whole grains, is thought to possibly help lessen insulin spikes, and therefore reduce the risk of type 2 diabetes. The link between increased fiber consumption and a decreased risk of colorectal cancer is still uncertain.

Fat

A molecule of dietary fat typically consists of several fatty acids (containing long chains of carbon and hydrogen atoms), bonded to a glycerol. They are typically found as triglycerides (three fatty acids attached to one glycerol backbone). Fats may be classified as saturated or unsaturated depending on the detailed structure of the fatty acids involved. Saturated fats have all of the carbon atoms in their fatty acid chains bonded to hydrogen atoms, whereas unsaturated fats have some of these carbon atoms double-bonded, so their molecules have relatively fewer hydrogen atoms than a saturated fatty acid of the same length. Unsaturated fats may be further classified as monounsaturated (one double-bond) or polyunsaturated (many double-bonds). Furthermore, depending on the location of the double-bond in the fatty acid chain, unsaturated fatty acids are classified as omega-3 or omega-6 fatty acids. Trans fats are a type of unsaturated fat with trans-isomer bonds; these are rare in nature and in foods from natural sources; they are typically created in an industrial process called (partial) hydrogenation. There are nine kilocalories in each gram of fat. Fatty acids such as conjugated linoleic acid, catalpic acid, eleostearic acid and punicic acid, in addition to providing energy, represent potent immune modulatory molecules.

Saturated fats (typically from animal sources) have been a staple in many world cultures for millennia. Unsaturated fats (e. g., vegetable oil) are considered healthier, while trans fats are to be avoided. Saturated and some trans fats are typically solid at room temperature (such as butter or lard), while unsaturated fats are typically liquids (such as olive oil or flaxseed oil). Trans fats are very rare in nature, and have been shown to be highly detrimental to human health, but have properties useful in the food processing industry, such as rancidity resistance.

Essential Fatty Acids

Most fatty acids are non-essential, meaning the body can produce them as needed, generally from other fatty acids and always by expending energy to do so. However, in humans, at least two fatty acids are essential and must be included in the diet. An appropriate balance of essential fatty acids—omega-3 and omega-6 fatty acids—seems also important for health, although definitive experimental demonstration has been elusive. Both of these "omega" long-chain polyunsaturated fatty acids are substrates for a class of eicosanoids known as prostaglandins, which have roles throughout the human body. They are hormones, in some respects. The omega-3 eicosapentaenoic acid (EPA), which can be made in the human body from the omega-3 essential fatty acid alpha-linolenic acid (ALA), or taken in through marine food sources, serves as a building block for series 3 prostaglandins (e.g. weakly inflammatory PGE3). The omega-6 dihomo-gamma-linolenic acid (DGLA) serves as a building block for series 1 prostaglandins (e.g. anti-inflammatory PGE1), whereas arachidonic acid (AA) serves as a building block for series 2 prostaglandins (e.g. pro-inflammatory PGE 2). Both DGLA and AA can be made from the omega-6 linoleic acid (LA) in the human body, or can be taken in directly through food. An appropriately balanced intake of omega-3 and omega-6 partly determines the relative production of different prostaglandins, which is one reason why a balance between omega-3 and omega-6 is believed important for cardiovascular health. In industrialized societies, people typically consume large amounts of processed vegetable oils, which have reduced amounts of the essential fatty acids along with too much of omega-6 fatty acids relative to omega-3 fatty acids.

The conversion rate of omega-6 DGLA to AA largely determines the production of the prostaglandins PGE1 and PGE2. Omega-3 EPA prevents AA from being released from membranes, thereby skewing prostaglandin balance away from pro-inflammatory PGE2 (made from AA) toward anti-inflammatory PGE1 (made from DGLA). Moreover, the conversion (desaturation) of DGLA to AA is controlled by the enzyme delta-5-desaturase, which in turn is controlled by hormones such as insulin (up-regulation) and glucagon (down-regulation). The amount and type of carbohydrates consumed, along with some types of amino acid, can influence processes involving insulin, glucagon, and other hormones; therefore the ratio of omega-3 versus omega-6 has wide effects on general health, and specific effects on immune function and inflammation, and mitosis (i.e. cell division).

Protein

Proteins are the basis of many animal body structures (e.g. muscles, skin, and hair). They also form the enzymes that control chemical reactions throughout the body. Each molecule is composed of amino acids, which are characterized by inclusion of nitrogen and sometimes sulphur (these components are responsible for the distinctive smell of burning protein, such as the keratin in hair). The body requires amino acids to produce new proteins (protein retention) and to replace damaged proteins (maintenance). As there is no protein or amino acid storage provision, amino acids must be present in the diet. Excess amino acids are discarded, typically in the urine. For all animals, some amino acids are essential (an animal cannot produce them internally) and some are non-essential (the animal can produce them from other nitrogen-containing compounds). About twenty amino acids are found in the human body, and about ten of these are essential and, therefore, must be included in the diet. A diet that contains adequate amounts of amino acids (especially those that are essential) is particularly important in some situations: during early development and maturation, pregnancy, lactation, or injury (a burn, for instance). A complete protein source contains all the essential amino acids; an incomplete protein source lacks one or more of the essential amino acids.

It is possible to combine two incomplete protein sources (e.g. rice and beans) to make a complete protein source, and characteristic combinations are the basis of distinct cultural cooking traditions. However, complementary sources of protein don't need to be eaten at the same meal to be used together by the body. Sources of dietary protein include meats, tofu and other soy-products, eggs, legumes, and dairy products such as milk and cheese. Excess amino acids from protein can be converted into glucose and used for fuel through a process called gluconeogenesis. The amino acids remaining after such conversion are discarded.

MINERALS

Dietary minerals are the chemical elements required by living organisms, other than the four elements carbon, hydrogen, nitrogen, and oxygen that are present in nearly all organic molecules. The term "mineral" is archaic, since the intent is to describe simply the less common elements in the diet. Some are heavier than the four just mentioned, including several metals, which often occur as ions in the body. Some dietitians recommend that these be supplied from foods in which they occur naturally, or at least as complex compounds, or sometimes even from natural inorganic sources (such as calcium carbonate from ground oyster shells). Some minerals are absorbed much more readily in the ionic forms found in such sources. On the other hand, minerals are often artificially added to the diet as supplements; the most famous is likely iodine in iodized salt which prevents goiter.

Macrominerals

Many elements are essential in relative quantity; they are usually called "bulk minerals". Some are structural, but many play a role as electrolytes. Elements with recommended dietary allowance (RDA) greater than 200 mg/day are, in alphabetical order (with informal or folk-medicine perspectives in parentheses):

- Calcium, a common electrolyte, but also needed structurally (for muscle and digestive system health, bone strength, some forms neutralize acidity, may help clear toxins, provides signaling ions for nerve and membrane functions)
- Magnesium, required for processing ATP and related reactions (builds bone, causes strong peristalsis, increases flexibility, increases alkalinity)
- Phosphorus, required component of bones; essential for energy

processing

- Potassium, a very common electrolyte (heart and nerve health)
- Sodium, a very common electrolyte; not generally found in dietary supplements, despite being needed in large quantities, because the ion is very common in food: typically as sodium chloride, or common salt. Excessive sodium consumption can deplete calcium and magnesium, leading to high blood pressure and osteoporosis.
- Sulfur, for three essential amino acids and therefore many proteins (skin, hair, nails, liver, and pancreas). Sulfur is not consumed alone, but in the form of sulfur-containing amino acids

Trace minerals

Many elements are required in trace amounts, usually because they play a catalytic role in enzymes. Some trace mineral elements (RDA < 200 mg/day) are, in alphabetical order:

- Cobalt required for biosynthesis of vitamin B12 family of coenzymes. Animals cannot biosynthesize B12, and must obtain this cobalt-containing vitamin in the diet
- Copper required component of many redox enzymes, including cytochrome c oxidase
- Chromium required for sugar metabolism
- Iodine required not only for the biosynthesis of thyroxine, but probably, for other important organs as breast, stomach, salivary glands, thymus etc. (see Extrathyroidal iodine); for this reason iodine is needed in larger quantities than others in this list, and sometimes classified with the macrominerals
- Iron required for many enzymes, and for hemoglobin and some other proteins
- Manganese (processing of oxygen)
- Molybdenum required for xanthine oxidase and related oxidases
- Nickel present in urease
- Selenium required for peroxidase (antioxidant proteins)
- Zinc required for several enzymes such as carboxypeptidase, liver alcohol dehydrogenase, and carbonic anhydrase

Vitamins

Some vitamins are recognized as essential nutrients, necessary in the diet for good health. (Vitamin D is the exception: it can be synthesized in the skin, in the presence of UVB radiation.) Certain vitamin-like compounds that are recommended in the diet, such as carnitine, are thought useful for survival and health, but these are not "essential" dietary nutrients because the human body has some capacity to produce them from other compounds. Moreover, thousands of different phytochemicals have recently been discovered in food (particularly in fresh vegetables), which may have desirable properties including antioxidant activity (see below); however, experimental demonstration has been suggestive but inconclusive. Other essential nutrients that are not classified as vitamins include essential amino acids (see above), choline, essential fatty acids (see above), and the minerals discussed in the preceding section.

Vitamin deficiencies may result in disease conditions, including goitre, scurvy, osteoporosis, impaired immune system, disorders of cell metabolism, certain forms of cancer, symptoms of premature aging, and poor psychological health (including eating disorders), among many others. Excess levels of some vitamins are also dangerous to health (notably vitamin A), and for at least one vitamin, B6, toxicity begins at levels not far above the required amount. Deficient or excess levels of minerals can also have serious health consequences.

Water

Water is excreted from the body in multiple forms; including urine and feces, sweating, and by water vapour in the exhaled breath. Therefore it is necessary to adequately rehydrate to replace lost fluids.

Early recommendations for the quantity of water required for maintenance of good health suggested that 6–8 glasses of water daily is the minimum to maintain proper hydration. However the notion that a person should consume eight glasses of water per day cannot be traced to a credible scientific source. The original water intake recommendation in 1945 by the Food and Nutrition Board of the National Research Council read: "An ordinary standard for diverse persons is 1 milliliter for each calorie of food. Most of this quantity is contained in prepared foods." More recent comparisons of well-known recommendations on fluid intake have revealed large discrepancies in the volumes of water we need to consume for good health. Therefore, to help standardize guidelines, recommendations for water consumption are included in two recent European Food Safety Authority (EFSA) documents (2010): (i) Food-based dietary guidelines and (ii) Dietary reference values for water or adequate daily intakes (ADI). These specifications were provided by calculating adequate intakes from measured intakes in populations of individuals with "desirable osmolarity values of urine and desirable water volumes per energy unit consumed." For healthful hydration, the current EFSA guidelines recommend total water intakes of 2.0 L/day for adult females and 2.5 L/day for adult males. These reference values include water from drinking water, other beverages, and from food. About 80% of our daily water requirement comes from the beverages we drink, with the remaining 20% coming from food. Water content varies depending on the type of food consumed, with fruit and vegetables containing more than cereals, for example. These values are estimated using country-specific food balance sheets published by the Food and Agriculture Organisation of the United Nations. Other guidelines for nutrition also have implications for the beverages we consume for healthy hydration- for example, the World Health Organization (WHO) recommend that added sugars should represent no more than 10% of total energy intake.

The EFSA panel also determined intakes for different populations. Recommended intake volumes in the elderly are the same as for adults as despite lower energy consumption, the water requirement of this group is increased due to a reduction in renal concentrating capacity. Pregnant and breastfeeding women require additional fluids to stay hydrated. The EFSA panel proposes that pregnant women should consume the same volume of water as non-pregnant women, plus an increase in proportion to the higher energy requirement, equal to 300 mL/day. To compensate for additional fluid output, breastfeeding women require an additional 700 mL/day above the recommended intake values for non-lactating women.

For those who have healthy kidneys, it is somewhat difficult to drink too much water, but (especially in warm humid weather and while exercising) it is dangerous to drink too little. While overhydration is much less common than dehydration, it is also possible to drink far more water than necessary which can result in water intoxication, a serious and potentially fatal condition. In particular, large amounts of de-ionized water are dangerous.

OTHER NUTRIENTS

Other micronutrients include antioxidants and phytochemicals. These substances are generally more recent discoveries that have not yet been recognized as vitamins or as required. Phytochemicals may act as antioxidants, but not all phytochemicals are antioxidants.

Antioxidants

As cellular metabolism/energy production requires oxygen, potentially damaging (e.g. mutation causing) compounds known as free radicals can form. Most of these are oxidizers (i.e. acceptors of electrons) and some react very strongly. For the continued normal cellular maintenance, growth, and division, these free radicals must be sufficiently neutralized by antioxidant compounds. Recently, some researchers suggested an interesting theory of evolution of dietary antioxidants. Some are produced by the human body with adequate precursors (glutathione, Vitamin C), and those the body cannot produce may only be obtained in the diet via direct sources (Vitamin C in humans, Vitamin A, Vitamin K) or produced by the body from other compounds (Beta-carotene converted to Vitamin A by the body, Vitamin D synthesized from cholesterol by sunlight). Phytochemicals (Section Below) and their subgroup, polyphenols, make up the majority of antioxidants; about 4,000 are known.

Different antioxidants are now known to function in a cooperative network. For example, Vitamin C can reactivate free radical-containing glutathione or Vitamin E by accepting the free radical itself. Some antioxidants are more effective than others at neutralizing different free radicals. Some cannot neutralize certain free radicals. Some cannot be present in certain areas of free radical development (Vitamin A is fat-soluble and protects fat areas, Vitamin C is water soluble and protects those areas). When interacting with a free radical, some antioxidants produce a different free radical compound that is less dangerous or more dangerous than the previous compound. Having a variety of antioxidants allows any byproducts to be safely dealt with by more efficient antioxidants in neutralizing a free radical's butterfly effect.

Although initial studies suggested that antioxidant supplements might promote health, later large clinical trials did not detect any benefit and suggested instead that excess supplementation may be harmful.

Phytochemicals

A growing area of interest is the effect upon human health of trace chemicals, collectively called phytochemicals. These nutrients are typically found in edible plants, especially colorful fruits and vegetables, but also other organisms including seafood, algae, and fungi. The effects of phytochemicals increasingly survive rigorous testing by prominent health organizations. One of the principal classes of phytochemicals are polyphenol antioxidants, chemicals that are known to provide certain health benefits to the cardiovascular system and immune system. These chemicals are known to down-regulate the formation of reactive oxygen species, key chemicals in cardiovascular disease.

Perhaps the most rigorously tested phytochemical is zeaxanthin, a yellow-pigmented carotenoid present in many yellow and orange fruits and vegetables. Repeated studies have shown a strong correlation between ingestion of zeaxanthin and the prevention and treatment of age-related macular degeneration (AMD). Less rigorous studies have proposed a correlation between zeaxanthin intake and cataracts. A second carotenoid, lutein, has also been shown to lower the risk of contracting AMD. Both compounds have been observed to collect in the retina when ingested orally, and they serve to protect the rods and cones against the destructive effects of light.

Another carotenoid, beta-cryptoxanthin, appears to protect against chronic joint inflammatory diseases, such as arthritis. While the association between serum blood levels of beta-cryptoxanthin and substantially decreased joint disease has been established, neither a convincing mechanism for such protection nor a cause-and-effect have been rigorously studied. Similarly, a red phytochemical, lycopene, has substantial credible evidence of negative association with development of prostate cancer.

Some of the correlations between the ingestion of certain phytochemicals and the prevention of disease are, in some cases, enormous in magnitude. Yet, even when the evidence is obtained, translating it to practical dietary advice can be difficult and counter-intuitive. Lutein, for example, occurs in many yellow and orange fruits and vegetables and protects the eyes against various diseases. However, it does not protect the eye nearly as well as zeaxanthin, and the presence of lutein in the retina will prevent zeaxanthin uptake. Additionally, evidence has shown that the lutein present in egg yolk is more readily absorbed than the lutein from vegetable sources, possibly because of fat solubility. At the most basic level, the question "should you eat eggs?" is complex to the point of dismay, including misperceptions about the health effects of cholesterol in egg yolk, and its saturated fat content.

As another example, lycopene is prevalent in tomatoes (and actually is the chemical that gives tomatoes their red color). It is more highly concentrated, however, in processed tomato products such as commercial pasta sauce, or tomato soup, than in fresh "healthy" tomatoes. Yet, such sauces tend to have high amounts of salt, sugar, other substances a person may wish or even need to avoid.

Intestinal Bacterial Flora

It is now also known that animal intestines contain a large population of gut flora. In humans, these include species such as Bacteroides, L. acidophilus and E. coli, among many others. They are essential to digestion, and are also affected by the food we eat. Bacteria in the gut perform many important functions for humans, including breaking down and aiding in the absorption of otherwise indigestible food; stimulating cell growth; repressing the growth of harmful bacteria, training the immune system to respond only to pathogens; producing vitamin B12, and defending against some infectious diseases.

Advice and Guidance

In the US, dietitians are registered (RD) or licensed (LD) with the Commission for Dietetic Registration and the American Dietetic Association, and are only able to use the title "dietitian," as described by the business and professions codes of each respective state, when they have met specific educational and experiential prerequisites and passed a national registration or licensure examination, respectively. In California, registered dietitians must abide by the "Business and Professions Code of Section 2585-2586.8".Anyone may call themselves a nutritionist, including unqualified dietitians, as this term is unregulated. Some states, such as the State of Florida, have begun to include the title "nutritionist" in state licensure requirements. Most governments provide guidance on nutrition, and some also impose mandatory disclosure/labeling requirements for processed food manufacturers and restaurants to assist consumers in complying with such guidance.In the US, nutritional standards and recommendations are established jointly by the US Department of Agriculture and US Department of Health and Human Services. Dietary and physical activity guidelines from the USDA are presented in the concept of myPlate, which superseded food pyramid, which superseded the Four Food Groups. The Senate committee currently responsible for oversight of the USDA is the Agriculture, Nutrition and Forestry Committee. Committee hearings are often televised on C-SPAN as seen here.

The U.S. Department of Health and Human Services provides a sample week-long menu which fulfills the nutritional recommendations of the government.

Government Programs

Federal and state governmental organizations have been working on nutrition literacy interventions in non-primary health care settings to address the nutrition information problem in the U.S. Some programs include:

The Family Nutrition Program (FNP) is a free nutrition education program serving low-income adults around the U.S. This program is funded by the Food Nutrition Service's (FNS) branch of the United States Department of Agriculture (USDA) usually through a local state academic institution which runs the program. The FNP has developed a series of tools to help families participating in the Food Stamp Program stretch their food dollar and form healthful eating habits including nutrition education.

Expanded Food and Nutrition Education Program (ENFEP) is a unique program that currently operates in all 50 states and in American Samoa, Guam, Micronesia, Northern Marianas, Puerto Rico, and the Virgin Islands. It is designed to assist limited-resource audiences in acquiring the knowledge, skills, attitudes, and changed behavior necessary for nutritionally sound diets, and to contribute to their personal development and the improvement of the total family diet and nutritional well-being.

An example of a state initiative to promote nutrition literacy is Smart Bodies, a public-private partnership between the state's largest university system and largest health insurer, Louisiana State Agricultural Center and Blue Cross and Blue Shield of Louisiana Foundation. Launched in 2005, this program promotes lifelong healthful eating patterns and physically active lifestyles for children and their families. It is an interactive educational program designed to help prevent childhood obesity through classroom activities that teach children healthful eating habits and physical exercise.

Teaching

Nutrition is taught in schools in many countries. In England and Wales the Personal and Social Education and Food Technology curricula include nutrition, stressing the importance of a balanced diet and teaching how to read nutrition labels on packaging. In many schools a Nutrition class will fall within the Family and Consumer Science or Health departments. In some American schools, students are required to take a certain number of FCS or Health related classes. Nutrition is offered at many schools, and if it is not a class of its own, nutrition is included in other FCS or Health classes such as: Life Skills, Independent Living, Single Survival, Freshmen Connection, Health etc. In many Nutrition classes, students learn about the food groups, the food pyramid, Daily Recommended Allowances, calories, vitamins, minerals, malnutrition, physical activity, healthful food choices and how to live a healthy life.

A 1985 US National Research Council report entitled Nutrition Education in US Medical Schools concluded that nutrition education in medical schools was inadequate. Only 20% of the schools surveyed taught nutrition as a separate, required course. A 2006 survey found that this number had risen to 30%.

Healthy Diets

Whole Plant Food Fiet

Heart disease, cancer, obesity, and diabetes are commonly called "Western" diseases because these maladies were once rarely seen in developing countries. An international study in China found some regions had essentially no cancer or heart disease, while in other areas they reflected "up to a 100-fold increase" coincident with shifts from diets that were found to be entirely plant-based to heavily animal-based, respectively. In contrast, diseases of affluence like cancer and heart disease are common throughout the developed world, including the United States. Adjusted for age and exercise, large regional clusters of people in China rarely suffered from these "Western" diseases possibly because their diets are rich in vegetables, fruits and whole grains, and have little dairy and meat products. Some studies show these to be, in high quantities, possible causes of some cancers. There are arguments for and against this controversial issue.

The United Healthcare/Pacificare nutrition guideline recommends a whole plant food diet, and recommends using protein only as a condiment with meals. A National Geographic cover article from November 2005, entitled The Secrets of Living Longer, also recommends a whole plant food diet. The article is a lifestyle survey of three populations, Sardinians, Okinawans, and Adventists, who generally display longevity and "suffer a fraction of the diseases that commonly kill people in other parts of the developed world, and enjoy more healthy years of life." In sum, they offer three sets of 'best practices' to emulate. The rest is up to you. In common with all three groups is to "Eat fruits, vegetables, and whole grains."

The National Geographic article noted that an NIH funded study of 34,000 Seventh-day Adventists between 1976 and 1988 "...found that the Adventists' habit of consuming beans, soy milk, tomatoes, and other fruits lowered their risk of developing certain cancers. It also suggested that eating whole grain bread, drinking five glasses of water a day, and, most surprisingly, consuming four servings of nuts a week reduced their risk of heart disease."

The French "paradox"

The French paradox is the observation that the French suffer a relatively low incidence of coronary heart disease, despite having a diet relatively rich in saturated fats. A number of explanations have been suggested:

— Saturated fat consumption does not cause heart disease
— Reduced consumption of processed carbohydrate and other junk foods.
— Regular consumption of red wine.
— More active lifestyles involving plenty of daily exercise, especially walking; the French are much less dependent on cars than Americans are.
— Higher consumption of artificially produced trans-fats by Americans, which has been shown to have greater lipoprotein effects per gram than saturated fat.

However, statistics collected by the World Health Organization from 1990–2000 show that the incidence of heart disease in France may have been underestimated and, in fact, may be similar to that of neighboring countries.

References

Corbridge D. E. C. (1995). *Phosphorus: An Outline of its Chemistry, Biochemistry, and Technology* (5th ed.). Amsterdam: Elsevier.

Lippard, S. J. and Berg, J. M. (1994). *Principles of Bioinorganic Chemistry*. Mill Valley, CA: University Science Books.

Nelson, D. L.; Cox, M. M. (2000). *Lehninger Principles of Biochemistry* (3rd ed.). New York: Worth Publishing.

Shils et al. (2005). *Modern Nutrition in Health and Disease*. Lippincott Williams and Wilkins.

WHO Technical Report Series. Diet, nutrition and the prevention of chronic diseases." Report of a Joint WHO/FAO Expert Consultation; Geneva 2003. Retrieved 2011-03-07

3

Healthy Diet

A healthy diet is one that helps maintain or improve general health. It is important for lowering many chronic health risks, such as obesity, heart disease, diabetes, hypertension and cancer. A healthy diet involves consuming appropriate amounts of all essential nutrients and an adequate amount of water. Nutrients can be obtained from many different foods, so there are numerous diets that may be considered healthy. A healthy diet needs to have a balance of fats, proteins, and carbohydrates, calories to support energy need and micro nutrients to meet the needs for human nutrition without inducing toxicity or excessive weight gain from consuming excessive amounts.

The World Health Organization (WHO) makes the following 5 recommendations with respect to both populations and individuals:

— Try to burn as much energy as you eat, and try to eat as much energy as you burn, as a healthy weight is a balance between those two.

— Increase consumption of plant foods, particularly fruits, vegetables, legumes, whole grains and nuts

— Limit intake of fat and oil, and avoid saturated fats, which are those that become solid at room temperature such as coconut oil and most animal fats including those found in red meat, dairy and eggs. Prefer unsaturated fats, which remain liquid at room temperature, and are predominant in most plant-based oils and foods. Eliminate trans fats.

— Limit the intake of granulated sugar — A 2003 report recommends less than 10% simple sugars

— Limit salt / sodium consumption from all sources and ensure that salt is iodized

Other recommendations include:

— Sufficient essential amino acids to provide cellular replenishment and transport proteins. All essential amino acids are present together only in animals. Many plants such as quinoa, soy, and hemp also provide almost all the essential acids but are always lacking in one or more essential amino acid. Those who omit meat products from their diets may still easily obtain all the essential amino acid in their diets by consuming adequate amounts of plant products and grain products, which combine to provide complete proteins to the diet. Fruits such as avocado and pumpkin seeds also have almost all of the essential amino acids.

— Essential micronutrients such as vitamins and certain minerals.

— Avoiding directly poisonous (e.g. heavy metals) and carcinogenic (e.g. benzene) substances;

— Avoiding foods contaminated by human pathogens (e.g. E. coli, tapeworm eggs).

Each basic food group is the major contributor of at least one nutrient while making substantial contributions of many other nutrients. Because each food group provides a wide array of nutrients in substantial amounts, it is important to include all food groups in the daily diet. Both illustrative eating patterns include a variety of nutrientdense foods within the major food groups. Selecting a variety of foods within the grain, vegetable, fruit, and meat groups may help to ensure that an adequate amount of nutrients and other potentially beneficial substances are consumed. For example, fish contains varying amounts of fatty acids that may be beneficial in reducing cardiovascular disease risk.

Nutrient Dense Foods

Nutrientdense foods are those foods that provide substantial amounts of vitamins and minerals (micronutrients) and relatively few calories. Foods that are low in nutrient density are foods that supply calories but relatively small amounts of micronutrients, sometimes none at all. The greater the consumption of foods or beverages that are low in nutrient density, the more difficult it is to consume enough nutrients without gaining weight, especially for sedentary individuals. The consumption of added sugars, saturated and trans fats, and alcohol provides calories while providing little, if any, of the essential nutrients.

Selecting lowfat forms of foods in each group and forms free of added sugars-in other words nutrientdense versions of foods-provides individuals a way to meet their nutrient needs while avoiding the overconsumption of calories and of food components such as saturated fats. However, Americans generally do not eat nutrientdense forms of foods. Most people will exceed calorie recommendations if they consistently choose higher fat foods within the food groups-even if they do not have dessert, sweetened beverages, or alcoholic beverages. If only nutrientdense foods are selected from each food group in the amounts proposed, a small amount of calories can be consumed as added fats or sugars, alcohol, or other foods-the discretionary calorie allowance.

The actual prevalence of inadequacy for a nutrient can be determined only if an Estimated Average Requirement (EAR) has been established and the distribution of usual dietary intake can be obtained. If such data are not available for a nutrient but there is evidence for a public health problem associated with low intakes, a nutrient might still be considered to be of concern. Based on these considerations, dietary intakes of the following nutrients may be low enough to be of concern for:

— *Adults:* calcium, potassium, fiber, magnesium, and vitamins A (as carotenoids), C, and E,

— *Children and adolescents:* calcium, potassium, fiber, magnesium, and vitamin E,

— *Specific population groups:* vitamin B_{12}, iron, folic acid, and vitamins E and D.

Efforts may be warranted to promote increased dietary intakes of potassium, fiber, and possibly vitamin E, regardless of age; increased intakes of calcium and possibly vitamins A (as carotenoids) and C and magnesium by adults; efforts are warranted to increase intakes of calcium and possibly magnesium by children age 9 years or older . Efforts may be especially warranted to improve the dietary intakes of adolescent females in general. Low intakes of fiber tend to reflect low intakes of whole grains, fruits, and vegetables. Low intakes of calcium tend to reflect low intakes of milk and milk products. Low intakes of vitamins A (as carotenoids) and C and magnesium tend to reflect low intakes of fruits and vegetables.

Selecting fruits, vegetables, whole grains, and lowfat and fatfree milk and milk products in the amounts suggested by the USDA Food Guide and the DASH Eating Plan will provide adequate amounts of these nutrients. Most Americans of all ages also need to increase their potassium intake. Most Americans may need to increase their consumption of foods rich in vitamin E (átocopherol) while decreasing their intake of foods high in energy but low in nutrients.

The vitamin E content in both the USDA Food Guide and the DASH Eating Plan is greater than current consumption, and specific vitamin Erich foods need to be included in the eating patterns to meet the recommended intake of vitamin E. Breakfast cereal that is fortified with vitamin E is an option for individuals seeking to increase their vitamin E intake while consuming a lowfat diet. In addition, most Americans need to decrease sodium intake. The DASH Eating Plan provides guidance on how to keep sodium intakes within recommendations.

People Over 50 and Vitamin B_{12}

Although a substantial proportion of individuals over age 50 have reduced ability to absorb naturally occurring vitamin B_{12}, they are able to absorb the crystalline form. Thus, all individuals over the age of 50 should be encouraged to meet their Recommended Dietary Allowance (RDA) (2.4 µg/day) for vitamin B_{12} by eating foods fortified with vitamin B_{12} such as fortified cereals, or by taking the crystalline form of vitamin B_{12} supplements.

Women and Iron

Based on blood values, substantial numbers of adolescent females and women of childbearing age are iron deficient. Thus, these groups should eat foods high in hemeiron (e.g., meats) and/or consume ironrich plant foods (e.g., spinach) or ironfortified foods with an enhancer of iron absorption, such as foods rich in vitamin C (e .g., orange juice).

Women and Folic Acid

Since folic acid reduces the risk of the neural tube defects, spina bifida, and anencephaly, a daily intake of 400 µg/day of synthetic folic acid (from fortified foods or supplements in addition to food forms of folate from a varied diet) is recommended for women of childbearing age who may become pregnant. Pregnant women should consume 600 µg/day of synthetic folic acid (from fortified foods or supplements) in addition to food forms of folate from a varied diet. It is not known whether the same level of protection could be achieved by using food that is naturally rich in folate.

Special groups and Vitamin D

Adequate vitamin D status, which depends on dietary intake and cutaneous synthesis, is important for optimal calcium absorption, and it can reduce the risk for bone loss. Two functionally relevant measures indicate that optimal serum 25hydroxyvitamin D may be as high as 80 nmol/L. The elderly and individuals with dark skin (because the ability to synthesise vitamin D from exposure to sunlight varies with degree of skin pigmentation) are at a greater risk of low serum 25hydroxyvitamin D concentrations.

Also at risk are those exposed to insufficient ultraviolet radiation (i.e., sunlight) for the cutaneous production of vitamin D (e.g., housebound individuals). For individuals within the highrisk groups, substantially higher daily intakes of vitamin D (i.e., 25 μg or 1,000 International Units (IU) of vitamin D per day) have been recommended to reach and maintain serum 25hydroxyvit -amin D values at 80 nmol/L. Three cups of vitamin Dfortified milk (7.5 μg or 300 IU), 1 cup of vitamin Dfortified orange juice (2.5 μg or 100 IU), and 15 μg (600 IU) of supplemental vitamin D would provide 25 μg (1,000 IU) of vitamin D daily.

Fluid

The combination of thirst and normal drinking behavior, especially the consumption of fluids with meals, is usually sufficient to maintain normal hydration. Healthy individuals who have routine access to fluids and who are not exposed to heat stress consume adequate water to meet their needs. Purposeful drinking is warranted for individuals who are exposed to heat stress or perform sustained vigorous activity.

Individual and Cultural Food Preferences

The USDA Food Guide and the DASH Eating Plan are flexible to permit food choices based on individual and cultural food preferences, cost, and availability. Both can also accommodate varied types of cuisines and special needs due to common food allergies. Two adaptations of the USDA Food Guide and the DASH Eating Plan are:

Vegetarians

Vegetarians of all types can achieve recommended nutrient intakes through careful selection of foods. These individuals should give special attention to their intakes of protein, iron, and vitamin B_{12}, as well as calcium and vitamin D if avoiding milk products. I n addition, vegetarians could select only nuts, seeds, and legumes from the meat and beans group, or they could include eggs if so desired. At the 2,000 calorie level, they could choose about 1.5 ounces of nuts and 2/3 cup legumes instead of 5.5 ounces of meat, poultry, and/or fish. One egg, ½ ounce of nuts, or ¼ cup of legumes is considered equivalent to 1 ounce of meat, poultry, or fish in the USDA Food Guide.

Milk and milk products

Since milk and milk products provide more than 70 percent of the calcium consumed by Americans, guidance on other choices of dietary calcium is needed for those who do not consume the recommended amount of milk products. Milk product consumption has been associated with overall diet quality and adequacy of intake of many nutrients, including calcium, potassium, magnesium, zinc, iron, riboflavin, vitamin A, folate, and vitamin D. People may avoid milk products because of allergies, cultural practices, taste, or other reasons. Those who avoid all milk products need to choose rich sources of the nutrients provided by milk, including potassium, vitamin A, and magnesium in addition to calcium and vitamin D.

The bioavailability of the calcium in these foods varies. Those who avoid milk because of its lactose content may obtain all the nutrients provided by the milk group by using lactosereduced or lowlactose milk products, taking small servings of milk several times a day, taking the enzyme lactase before consuming milk products, or eating other calciumrich foods.

Components of Healthy Diet

The number of studies have shown that we are cutting our consumption of high-fat meats and whole-milk dairy products, while increasing the amount of fish, chicken, pasta, grains, and fresh fruits and vegetables we eat. Food manufacturers are beginning to respond: the array of low-fat, or even nonfat, and low-sodium products has burgeoned; what's more, many have been reformulated as food science has become more sophisticated. So-called 'healthy' foods now taste better than they did as recently as a year or two ago. Even fast-food restaurant chains have reacted, replacing animal fats with vegetable fats in some of their fried foods, cutting the percentage of fat in their hamburgers, offering grilled as well as deep-fried chicken, and opening salad bars.

Animal products, which were the source of 70 percent of fat consumed in 1960, were only 57 percent in 1982, a reflection in part of the public's switch to margarine in place of butter and the disappearance of lard in many commercial and homemade baked goods. Even so, those who follow food trends note that sales of premium ice creams and high-fat cheeses have increased exponentially, as has consumption of soft drinks and sweets. Restaurateurs find that even those who conscientiously order fish and steamed vegetables for dinner reward themselves by having the fudge pie for dessert.

About half of all Americans are overweight. There are vast numbers of people who haven't yet been persuaded to change their eating habits, and legions more who are confused by seemingly conflicting claims ("Oat bran lowers cholesterol" vs. "Oat bran is no better than other fibers"), or who want to change but don't know how to begin. Average intake of saturated fat is at least 50 percent higher than most experts believe it should be, while cholesterol and sodium intake levels are twice the recommended levels.

The total number of calories in the diet, as well as the proportion of those calories that comes from fat, has been associated with coronary artery disease. Caloric intake that exceeds the amount needed to fuel the body's activities results in excess weight. An individual whose weight exceeds the desirable level by 20 percent or more is considered obese, and obesity is an independent risk factor for heart disease. In addition, it affects blood pressure and blood cholesterol and triglyceride levels and contributes to the development of diabetes.

The energy the body needs to carry out its basic functions comes from three sources: carbohydrates, proteins, and fats. Most foods area mixture of more than one energy source. Meat and cheese, for example, are considered protein foods, but can have large amounts of fat. Except for processed meats, such as luncheon meats, these animal products have no carbohydrates. (Milk is the only animal product that contains a significant amount of carbohydrates, in the form of lactose, a type of sugar.)

Foods generally classified as carbohydrate foods, which are plant products, can have high amounts of protein, are usually very low in fat, and rarely contain saturated fats. Fats, while found in other foods, are most familiar in their pure forms, such as butter, margarine, and oil. When foods containing these three basic elements are combined in appropriate proportions and in a variety of forms, they make up a balanced diet that provides the body with all the vitamins and minerals it needs and a sufficient amount of energy to function.

Carbohydrates

Carbohydrates are the body's major energy source, fuelling the activities of the brain, central nervous system, and muscles. What the body doesn't use immediately as glucose, it converts and stores as glycogen and fat. Glycogen, which is stored in the liver and the muscles, can be broken down quickly to restore blood glucose levels rapidly when they drop or to provide fuel for exercise. The amount of glycogen that can be stored is limited; the amount of fat, unfortunately, is not.

There are three main types of carbohydrates:

— starches (also called complex carbohydrates), which are found primarily in cereals, grains, and starchy vegetables;
— sugars (also called simple carbohydrates), which occur naturally in fruits, berries, and some vegetables, and are also found in refined form as table sugar and syrup and in processed foods; and
— fibers, which are found in whole grains, beans, legumes, fruits, and vegetables.

Sugar is best obtained from fruits and certain vegetables, because they provide vitamins, minerals, and fiber, rather than from foods high in refined sugars (such as commercial baked goods), which add nothing but calories and may also be high in fat. While there is no evidence that too much sugar, refined or otherwise, will lead to high blood sugar (hyperglycemia), glucose intolerance, or diabetes, it may contribute to obesity, which is a risk factor for diabetes and heart disease.

Starches, having suffered for many years from an image problem, are finally being recognised as mainstays of a healthy diet. No longer considered lowly peasant food (or worse, fattening!), they now appear in the trendiest restaurants in innovative dishes drawn from a multitude of cuisines. The public is coming to recognise that potatoes, breads, beans, rice dishes, and pastas are actually relatively low in calories for the amount of satiety (the feeling of fullness) they offer. Their poor reputation has come from the fact that we have traditionally prepared and eaten them with butter, cream sauces, and cheese.

Complex carbohydrates should be represented in the diet in more servings per day-6 to n-than any other category of food. Fortunately, they come in great variety and tend to be delicate in taste, so they lend themselves to the addition of herbs, spices, and other ingredients that heighten flavour and texture. Fiber, or roughage, is a type of carbohydrate that is not broken down by the body during digestion. It can be divided into two types, soluble and insoluble, and a healthy diet should have some of each.

Soluble fibers-such as pectin, guar (a common thickening agent), and the fibers found in barley, oat bran, legumes, and dried beans-are those that mix with water to become a gummy gel. Research suggests that soluble fiber may help lower the level of cholesterol, especially low-density-lipoprotein (LDL) cholesterol, in the blood. A diet that is high in complex carbohydrates and fiber and in which about one-third of the fiber is soluble has been shown in several studies to lower insulin requirements and improve blood-sugar control in some diabetics.

Insoluble fibers are the parts of plants that do not dissolve in water, but pass through the digestive system essentially intact. In fact, they absorb water and thus produce a soft, bulky stool. These fibers (cellulose is the most common kind) are found in most grains and seeds and in the skins of fruits and vegetables. High-fiber foods typically are low in calories and fat and thus beneficial to people trying to control their weight or their blood cholesterol levels. Noting that populations that eat a diet high in insoluble fiber have low rates of colon cancer, some scientists believe that insoluble fiber may play a role in reducing the risk of this type of cancer. Although there is no consensus on this, there is agreement that insoluble fiber helps prevent constipation and, by promoting a feeling of fullness, may aid in weight control.

Protein

Protein is the key element in the body's metabolic processes. It is vital to the growth and maintenance of tissues that make-up the brain, muscles, connective tissue, skin, hair, blood, and organs, and to the production of infection-fighting cells and antibodies. Because protein is so essential to life, the body has developed a unique system for recycling it. The proteins in the body are constantly being turned over and broken down, and the end products excreted. When protein is consumed in the proper amounts it is broken down by the body to provide new building blocks, called amino acids, for the body's own proteins. However, when more protein is eaten than is necessary for this replacement process, the excess amino acids are used as fuel or stored as fat.

There are 20 kinds of amino acids, and the body can produce 11 of them. The remaining nine, known as the "essential" amino acids, must be obtained from food. Most of the protein from animal sources-meat, poultry, fish, and diary products-is "complete"-that is, it contains all nine essential amino acids in the proper proportions. In contrast, vegetable sources of protein are incomplete-they lack certain essential amino acids. Strict vegetarians can nevertheless get the full complement of amino acids by combining legumes with grains, nuts, or seeds. Rice with beans and pasta with peas are good combinations.

It was once assumed that because protein is so vital to supporting the maintenance of the body, the higher the protein content, the better the diet. Now, however, it is recognised that excess protein is not risk-free. A high-protein diet makes the kidney work harder to excrete urea, the end product of proteinamino-acid breakdown. The kidneys work at full capacity after a high-protein meal, which a normal kidney can handle without difficulty. People with diabetes mellitus, especially insulin-dependent diabetics, are prone to kidney disease, and the progression of this disease may be slowed by a diet very low in protein.

Fat

Although fats have taken on a villainous role in the minds of many people, they are essential in small amounts to promote growth in children, for energy reserves, to carry vitamins A, D, E, and K in the bloodstream, bloodstream, and to manufacture prostaglandins, sex hormones, and cell membranes. They also keep skin from getting too dry, and in food they add flavour, texture, and aroma. They help promote a feeling of fullness (although the reason for this is not clearly understood) and keep the feeling of hunger from returning as quickly as it does after protein and carbohydrate meals.

Fats are part of the broad category of lipids, which includes fatty acids, triglycerides, sterols, cholesterol, and other substances not soluble in water. Dietary fats, as well as oils (which are a form of fat), can be classified as saturated, nonsaturated, or polyunsaturated. Most fat-containing foods, even those such as oil that are 100 percent fat, contain a mixture of these three types.

Saturated fat interferes with the removal of cholesterol from the blood and thus has the effect of raising the levels of serum cholesterol-the cholesterol circulating in the bloodstream. This happens because saturated fat appears to curtail the entry of the cholesterol-carrying low-density lipoproteins (LDLs) into the cells. When they do not get enough LDL cholesterol, the cells (especially those in the liver) make their own. The excess production is returned to the bloodstream, raising serum cholesterol levels.

Saturated fats, which can also serve as a precursor to cholesterol, have a greater effect on raising serum cholesterol than dietary cholesterol itself. In contrast, monounsaturated and polyunsaturated fats help lower the amount of total cholesterol in the blood by lowering LDL-cholesterol, although polyunsaturates also seem to lower HDL (the good cholesterol). With the exception of the tropical oils-palm, palm kernel, and coconut-saturated fats come from animal sources and tend to be solid at room temperature. The most familiar sources are butter, cheese, and the fat found in red meats.

Monounsaturated oils are liquid at room temperature, although they may become solid in the refrigerator. Major sources of monounsaturates include canola, olive, peanut and cashew oils, as well as olives, cashews, peanuts, and peanut products. Polyunsaturated oils, which are liquid even at cold temperatures, include cottonseed, safflower, sunflower, soybean, and corn oil, as well as the oils found in almonds, filberts, and pecans.

Hydrogenation, commonly used in the manufacture of margarine, is a chemical process used to make liquid fats more solid. In doing so, it makes them more saturated. For this reason, scientists once believed that margarine containing large amounts of hydrogenated oils were not as beneficial as softer margar-ines. As analysis of fatty acids has become more sophisticated, researchers have come to realise that hydrogenation changes some oils, such as corn oil, to stearic acid, the one form of saturated fat that does not raise serum cholesterol. Thus, the type of oil used in margarine has become more important than the relative hardness or softness of the margarine. Corn, safflower, and soybean oil margarine are all good choices.

In recent years, Americans have been eating less saturated fat, but they are also consuming an unprecedented amount of polyunsaturated oils. Few populations have ever eaten as much polyunsaturated oil as we currently do. Some studies have raised concern that excessive consumption of polyunsaturates could suppress the immune system and raise the risks of cancer and gallstones, but this has not been proved. At present, the benefits of a diet higher in polyunsaturated fats than in saturated fats seem clearly to outweigh the risks. Nevertheless, other research has shown that monounsaturates maybe better than polyunsaturates in conserving "good" highdensity-lipoprotein (HDL) cholesterol while lowering low-density-lipoprotein (LDL) cholesterol.

Olive oil has traditionally been consumed in large quantities, without any known adverse effects, in Italy, Spain, and Greece, where death rates from coronary artery disease are lower than in the United States. However, that does not mean that oils high in monounsaturates should be used to excess. Because high total fat intake-not just excessive saturated fat-is a risk factor for coronary artery disease, the goal is to lower the total intake of fat, not simply to replace saturated fat with other forms of fat. Instead, complex carbohydrates are the best substitution for saturated fat in the diet.

Finally, fat seems to play a unique role in weight maintenance, regardless of calorie level. It was once believed that "a calorie is a calorie" and that the most important factor in weight reduction was the total amount of calories consumed. A number of studies now show that not all calories are created equal, at least as far as fat calories are concerned. A person will lose weight faster on a low-fat, high-carbohydrate diet than on a high-fat diet of the same number of calories. This seems to be because the body metabolises fat differently from protein and carbohydrates. Some studies, for example, have shown that about 25 percent of calories from carbohydrates are burned by the body in the process of converting them for storage as fat tissue. Only about 3 percent of fat calories are burned in the metabolic process that converts them to body fat, so more is left to store.

Yale Nutrition Services recommendations are based on the principles of variety, balance, and moderation. No individual food is prohibited, but some categories of food are stressed over others. By choosing a variety of foods in each category, individuals will be able to meet the recommended dietary allowances (RDAs) for vitamins and minerals without the need for supplements. Compared to the typical American intake, the recommended diet contains fewer calories and less total fat, saturated fat, cholesterol, salt, and alcohol, and slightly less protein. The percentage of complex carbohydrates and fiber is increased, with the addition of more fruits, vegetables, cereals, grains, and legumes.

The recommendations call for an overall daily intake measured in percentages of total calories as follows:

— At least 55 percent carbohydrates
— Less than 30 percent fat, including:
  — less than 10 percent saturated fat
  — 10-15 percent monounsaturated fat
  — up to 10 percent polyunsaturated fat
— 15-20 percent protein

In addition:

— Cholesterol intake should be no more than 250 milligrams a day.
— Sodium intake should be 1,000 to 3,000 milligrams a day.

Determination of Calorie in Diet

Implementing the diet requires doing some calculations. Although this may seem a complex process, it only has to be done once. Once the daily fat grams are determined, it's only a matter of keeping track of these. And after a short while on a low-fat diet, as an individual begins to develop a sense of how many grams are in various foods, even counting the grams becomes unnecessary.

Figuring Fat Grams

Once caloric intake has been determined, the next step is to figure the appropriate number of grams of fat to average per day. Fats, as well as proteins and carbohydrates, are usually listed in grams on packaged foods (there are about 28 grams to 1 ounce). There are about 4 calories in every gram of protein or carbohydrate, 7 in each gram of alcohol, and 9 in each gram of fat. So gram for gram, fats have more than twice the calories of carbohydrates or proteins. The percentage figures in the dietary recommendations are based solely on the caloric contributions of various foods, not their weight or volume. To understand the difference, consider a certain type of ham that claims to be 95 percent fat-free and has 128 calories in a serving of 100 grams (3.5 ounces).

Table 1. Maximum Daily Fat Intake by Calorie Level

Total calorie level [calories/day)	*Maximum total fat (grams/day]*	*Maximum saturated fat (grams/day)*
1,000	33	11
1,200	40	13
1,400	47	15
1,600	53	17
1,800	60	20
2,000	67	22
2,200	73	24
2,400	80	26
2,600	87	28
2,800	93	30
3,000	100	32

That means that in 100 grams it has 5 grams of fat and 95 grams of other ingredients (water, protein, carbohydrates). Thus the claim is true-only 5 percent of the weight comes from fat. But since each gram of fat is 9 calories, 45 of the 128 calories-or 35 percent-comes from fat.

Servings and Protein Sizes

Although ultimately individuals will need to consider their average fat-gram allowance in choosing what to eat, most people find it difficult to make the leap from fat grams to menus. An easier way to develop an eating plan is to think in terms of servings per day. Generally, to follow a low-fat, high-carbohydrate diet and still get sufficient calories and nutrients (vitamins and minerals) an individual should have the following each day

— Three to five servings of vegetables, including at least one leafy green vegetable and one yellow or orange one. (One serving is 1 cup raw or ½ cup cooked.)

— Two to four servings of fruit. (One serving equals one piece of medium-sized fresh fruit or ½ cup juice or canned fruit.)

— Six to 11 servings of breads, cereals, pasta, rice, dried peas or beans. (One serving equals a slice of bread, 1 cup dry cereal or popcorn, or ½ cup cooked cereal, pasta, rice, dried peas, or beans.)

— Two servings (three for pregnant or breastfeeding women) of skim or

1 percent fat milk and low-fat or nonfat dairy products. (One serving equals 8 ounces of milk or yogurt, 1 ounce of a hard cheese containing less than 5 grams of fat, 1/3 cup of cottage cheese.)

— Two servings or less of lean, well-trimmed beef, veal, lamb, pork, skinless poultry, or fish; poultry and fish should be favoured over meat. (One serving equals 3 ounces cooked weight.) Or substitute nuts, dried peas, beans, lentils (1 cup serving equals 2 to 3 ounces of meat, fish, or poultry) or peanut butter (4 tablespoons equal 2 ounces of meat, fish, or poultry).

— Five to eight servings of margarine or polyunsaturated or monounsaturated oils (one serving equals 1 teaspoon).

The following are exceptions:

— Egg whites are unlimited; egg yolks should be limited to three or four a week, including those used in cooking and baking.

— Shrimp and lobster are low in fat, but somewhat higher in cholesterol than other fish and shellfish, and should be limited to one serving a week.

— Organ meats are high in cholesterol and should be eaten only occasionally. Liver, however, is an excellent source of vitamins and iron (especially important for premenopausal women) and is recommended about once a month.

The spread in the number of servings accounts for a difference in total caloric intake. Thus, while everyone should have at least six servings from the bread and cereal group, 11 servings may only be appropriate for a 6-foot male in his 20s who is not obese. Some people, once they recognise that fish and chicken are more healthful than fatty red meat and that olive oil is better than butter, go overboard and begin to eat the "good" foods in unlimited quantities. Paying attention to the number of servings and portion sizes is crucial to having a healthful diet. The sizes of standard portions-particularly the meat servings-may come as a surprise to some people.

Three ounces (cooked weight) of meat is about the size of one's palm or a deck of playing cards, a good deal smaller than the T-bone steaks some restaurants routinely serve. On the other hand, some individuals may be daunted by the idea of three to five servings of vegetables. These add up quickly, however, as most people serve more than half a cup of an individual vegetable.

The recommended diet emphasises complex carbohydrates, fruits, and vegetables, which are rich in vitamins and minerals and high in fiber. Complex carbohydrates come in great variety and include starchy vegetables (potatoes, corn, and green peas), legumes, grains, and nuts and seeds. Legumes, in turn, consist of beans (black, cranberry, fava, kidney, lima, pinto, mung, navy, pea, and soy, among others), peas (blackeyed, chick, cow, field, and split), peanuts, and lentils. Grains include barley, corn, oats, rice, rye, wheat (including bulgur, wheat germ, and sprouts), as well as flours and cereals made from these grains. Nuts (almonds, Brazil nuts, cashews, filberts, pecans, and walnuts, among others) and seeds (pumpkin, sunflower, and sesame) are carbohydrates that, like legumes and grains, are high in protein. Because they contain more fat, nuts and seeds should be used in smaller quantities than the other complex carbohydrates.

Combining any of the legumes with any of the grains, nuts, or seeds produces the complete array of essential amino acids and thus provides all the protein necessary for a healthful diet. Alternatively, grains can be combined with nuts or seeds. Adding a small amount of a nonfat or low-fat dairy product (such as milk, cheese, or yogurt) to legumes, grains, nuts, or seeds will also satisfy protein requirements. Other good, relatively low-fat sources of protein include fish, poultry (except duck and goose) without its skin and fat, lean red meat (beef, veal, lamb, and pork), and low-fat or nonfat dairy products. Fish, poultry, or meatless meals should be substituted for meat several times a week.

The growing consumer market for fish has made it possible to find fresh or flash-frozen fish thousands of miles from their source. If possible, a variety of fish should be chosen. The more fatty fish, such as salmon, mackerel, herring, rainbow trout, whitefish, and striped bass, are high in omega-3 fatty acids, a form of polyunsaturated fat that is chemically different from the omega-6 fatty acids predominant in vegetable oils. Omega-3 fatty acids may lower blood levels of triglycerides and very-low density lipoproteins (VLDLs). Less oily, milder-tasting fish such as flounder, haddock, cod, and sole are much lower in omega-3 fatty acids, but they are still beneficial, because they are low in fat and cholesterol. Like oily fish, they are also high in protein, zinc, and B vitamins, especially niacin and B6 Shellfish, while somewhat higher in cholesterol than fin fish, are still moderate in cholesterol and low in fat and can be eaten once a week.

Although there are other arguments for eliminating meat from the diet, it is not necessary to give up red meat completely to maintain a low-fat eating plan. It is particularly important for children, teenagers, and women of child-bearing age to consume foods that are good sources of iron. These include fish, leafy green vegetables, and iron-enriched breads and cereals, but the most concentrated source is still lean red meat. It is possible, although it sometimes requires the use of supplements, to get sufficient iron without red meat, but even a few ounces of meat a week provide enough to meet iron requirements for these groups.

For those who want to continue eating meat, the best way to use it is as a side dish, rather than as the main component of a meal. For example, dinner might be couscous tossed with herbs, carrots, broccoli, zucchini, and mushroom and some julienned strips of lean beef. Lean cuts of beef include round tip, top loin, top round, eye of round, tenderloin, top sirloin, rump, and flank (trimmed). Lean pork cuts include tender-loin, sirloin roast, and loin chops. The leg is the lean-est cut of lamb, and all cuts of veal are low in fat except the breast.

Skim or 1 percent fat milk, buttermilk, skim milk cheeses and yogurt, and margarine should be used in place of whole milk, high-fat cheeses, cream, and butter. The number of new nonfat products-cottage cheese, yogurt, frozen yogurt, mock sour cream, and mozzarella and ricotta cheeses, among others-seems to grow by the month. These dairy products are also high in calcium, which is important for children, teenagers, pregnant and nursing women, and premenopausal women in general (to help prevent the bone-thinning disease osteoporosis).

Use of Beverages

Caffeine-containing beverages, especially coffee, have been the object of much study, and the results have been conflicting and confusing. Some of the studies showing that coffee drinking increases the risk of heart disease have turned out to be poorly designed. A careful, large-scale study of more than 45,500 men reported in the New England Journal of Medicine in 1990 now seems to have satisfied the critics. It found that coffee drinking, even at levels of 6 or more cups a day, was not associated with an increased risk of heart disease or stroke. In fact, the only increased risk seemed to be with 4 or more cups a day of decaffeinated coffee. In any case, most experts feel that the key for healthy adults without symptoms is still moderation-no more than 5 cups a day. For those in whom caffeine increases anxiety, exacerbates arrhythmias, or produces other undesirable symptoms, even 1 or 2 cups maybe too many.

Moderation is also the watchword for adults who drink alcoholic beverages. While light drinking may raise protective HDL cholesterol levels, excessive alcohol consumption can raise blood pressure and triglyceride levels. It can also cause a type of heart disease known as alcoholic cardiomyopathy. The generally accepted recommendation for those who drink is no more than 1 to 2 ounces of ethanol a day. This translates into about 2 glasses of beer or wine or 2 ounces of hard liquor. The other important reason to limit alcohol consumption is weight control. Each gram of alcohol has 7 calories (there are about 200 calories in an ounce), compared to 4 in carbohydrates or protein. More important, calories from alcohol are considered "empty calories," since they add no nutritive value to the diet. When alcohol is served in a mixed drink, such as a Manhattan or a whiskey sour, the calorie count is even higher.

Foods to useless frequently are those high in cholesterol, such as egg yolks and organ meats, and those high in saturated fat. The list of high-saturated-fat foods is long, but several categories stand out fried foods, luncheon meats (bologna, salami, liverwurst, etc.), rich, creamy desserts, and many commercial baked goods. Careful shoppers can now find some of the more traditional luncheon meats such as pastrami that are made with turkey instead of pork or beef, but some of these substitutes and virtually all of the originals are high in sodium. Also available are nonfat coffee cakes, nonfat frozen yogurts and low-fat versions of other creamy desserts, and cookies made with unsaturated oils. Many manufacturers have responded to public demand by dropping tropical oils, such as coconut and palm, from their desserts. Food labels that merely say "vegetable oils," however, are suspect, since this broad category includes tropical oils.

Weight Maintaining

Obesity (20 percent or more above ideal weight for a year or more) is a significant risk factor for coronary artery disease. It also raises the risk of a variety of other health problems, including diabetes, hypertension, and even arthritis and back pain.

When weight control experts talk about overweight, they are really talking about "overfat." Because muscle weighs more than fat, a physically fit person who has a greater percentage of body weight as muscle than a sedentary person does can actually weigh more but not be overweight. For this reason, weight tables, while convenient, are not as accurate an indication of whether a person is at his or her ideal weight as are other methods. The simplest of these is the mirror test-if the person looks flabby and fat, chances are he or she is overweight.

More elaborate tests include skinfold thickness (measuring body fat with calipers) and underwater weighing. Individuals can get a general sense of whether they are carrying excess body fat by measuring the fat on the upper arm. Hold the arm out at shoulder level, bend the elbow, and use the thumb and finger of the other hand to gently pinch the flesh on the underside of the upper arm midway between the shoulder and elbow. Remove finger and thumb, keeping them the same distance apart, and measure; anything more than an inch between them is too much.

How excess fat is distributed may also affect the risk of coronary heart disease and other obesityrelated diseases. The risk seems to be higher among those who have what is referred to as a male fatdistribution pattern, or an apple, rather than a pear, shape.

Apple-shaped people (primarily men and postmenopausal women) tend to carry their excess weight around their waist (the potbelly), while pearshaped people (primarily premenopausal women) carry their excess weight on the hips, buttocks, and thighs. One explanation for this may be that fat in the abdomen is metabolically more active than fat in other parts of the body.

Fatty acids released into the bloodstream when abdominal fat is metabolised find their way into the nearby portal vein, through which they are transported directly to the liver. This in turn stimulates increased cholesterol output. Hormones seem to control fat distribution, although the tendency toward one pattern or the other is inherited. When estrogen production declines after menopause, women tend to accumulate more abdominal fat and their risk of heart disease rises. A man is at increased risk of heart disease if his waist size exceeds his hip size; for a woman, risk increases if her waist measurement is more than 80 percent of her hip measurement. For example, a woman who has 40-inch hips should not have a waist measurement of more than 32 inches in order to be in the low-risk category. Although nothing can be done to change the inherited fat-distribution pattern, the weight itself can be lost through diet and exercise.

Body weight is governed by a simple equation: The excess calories in the food eaten, minus the calories burned during exercise, equal the extra weight that the body will accumulate. Exercise not only consumes calories in and of itself, it also revs up the body's metabolic rate so that calories are expended ("burned") at a higher rate, even at rest. This helps reverse the slowdown in the body's metabolic "burn rate" that can result from eating less. Not only does exercise aid in losing extra pounds and maintaining ideal weight, it also may bean independent protective factor against coronary artery disease.

People who are overweight despite eating a healthful, low-fat diet can achieve and maintain an appropriate weight level simply by exercising more and eating less of each food group, without eliminating anything. They can continue to eat the variety of foods they favoured all along, but in smaller amounts. The majority of people who are overweight, however, are still eating a diet with excess fat-especially saturated fat. Simply by concentrating on eating less fat, they will consume fewer total calories.

Losing weight is easier than keeping the weight off. All too often, people resort to their old behaviour (overeating and inactivity) and gain back their weight-and then some-after "dieting." The regained weight tends to represent excess body fat rather than muscle. When people allow their weight to go up and down like a yo-yo, they end up flabbier than when they started. Even people who are not chronic dieters may fall victim to this yo-yo syndrome, as winter inactivity and overeating often cause yearly weight cycling: up in the fall, down in the spring.

To lose weight and maintain an ideal level, individuals have to concentrate on the long haul. The most important factor in effecting a permanent change is changing behaviour with respect to food as well as activity. Favourite foods do not have to be eliminated entirely, even if they are high in fat. In fact, cutting out these foods entirely leads to cravings and often results in binges. Instead, the goal is to learn to enjoy them sparingly-for instance, savouring one chocolate chip cookie instead of devouring an entire jar of them at one sitting.

Although all the information needed for successful weight 10SS is contained here, some people find they do better in a structured programme where they get peer support or are accountable to a health professional or group leader. In choosing such a programme there are two red flags to watch for. One is a diet programme that does not emphasise the crucial role of regular exercise in weight control. The other is any regimen that seems monotonous and does not allow a variety of foods. Drastic, formula-type or short-term diets that are restricted to a narrow range of foods may produce immediate weight loss, but they are rarely successful in long-term weight management. In addition, such regimens promote the breakdown of lean muscle tissue, including heart muscle, at least far the first few days of the diet. This happens because it takes the body longer to mobilise its fat stores for energy than it does to cannibalise its own protein.

In addition, the body tries to compensate for a starvation diet by lowering its metabolic rate. Thus, after a low-calorie diet, the body can get by on fewer calories than before. For example, a person who previously needed 1,800 calories a day to maintain normal weight may now need only 1,500 calories after a short-term drastic diet. Consequently, this person will gain weight even if he or she consumes only 1,800 calories. These low-calorie diets also provoke a feeling of deprivation, and for that reason they almost invariably backfire. Even if people manage to take off their unwanted weight using such diets, they often reward themselves after the diet is over by overindulging in the formerly forbidden foods and regaining the shed pounds. Furthermore, many of these diets are so low in calories (below 1,200) that they cannot be nutritionally complete.A slow, steady weight loss-no more than 1 or 2 pounds a week-is easiest to maintain. In fact, we feel so strongly about this that when participants in our diet groups consistently lose more than 2 pounds a week, they are dropped from the programme because they are merely perpetuating unproductive behaviour. The best diet, then, for losing weight and maintaining the loss is one that can be enjoyed indefinitely.

Tips for Dietary Recommendations in Daily Routine

Practical Tips

Shopping for food

It remains generally true that people can achieve a healthier diet by buying fresh food, preparing it themselves, and avoiding prepackaged foods.

However, by shopping carefully and especially by reading labels, consumers can find healthful prepared foods. While some manufacturers have responded to public demand with excellent low-fat, nonfat, and low-salt products, others have attempted to exploit consumer interest with deceptive labelling practices. One common ploy, for example, is used by producers of some vegetable oils and peanut butters who proclaim that the products contain no cholesterol. The claim ignores the fact that cholesterol is found only in animal foods, not in vegetable products. The unwary consumer may be led to believe that cholesterol has been specially eliminated from these particular products. The problem is that health claims have become so widespread on product packaging that the uninformed consumer can easily be confused.

New regulations from the Food and Drug Administration (FDA), which take effect in 1993, should end many of these practices. Nutrition labelling will be required for most foods under the jurisdiction of the FDA, although not for foods, notably meat and poultry, that come under the jurisdiction of the U.S. Department of Agriculture. At present the FDA is still developing exact definitions for certain descriptive phrases such as "light" or "lite", setting standards for how nutrition information will be visually interpreted, and making decisions on other issues such as state and federal overlap of regulations.

There are certain standards that manufacturers must follow even now. All canned, frozen, and packaged products (except fresh foods) must list ingredients in descending order according to volume. The only exceptions are certain standardised products, such as ketchup, that meet "standards of identity' set by the FDA. All products containing fat must list the kind of animal or vegetable fat used. Finally, all enriched or fortified foods and those for which the manufacturer makes nutritional claims must carry a nutrition label containing certain facts about the product. These include serving size, number of servings per container, and the amount of carbohydrates, protein, fat, calories, and sodium per serving.

Saturated fat, polyunsaturated fat, cholesterol, and other items are optional unless the manufacturer makes a specific claim about them (such as that a product is low in cholesterol). Beginning in 1993, saturated fat, cholesterol, sodium, total carbohydrates, complex carbohydrates and sugar, total protein, and dietary fiber must also appear on the food labels. The serving size must be expressed in common household terms (such as ½ cup).

The key items to pay attention to are the serving size, calories, and grams of fat. Some serving sizes listed on labels are unrealistically small in order to make the fat or calories appear low; most are correct, but some are smaller than many people are used to eating (2 teaspoons of peanut butter, for example). Consumers need to keep the serving size in mind when eating, but the purchase decision can be made on the basis of a simple rule of thumb: 3 grams of fat per 100 calories a serving. At 9 calories a gram, 3 grams would be about 27 calories, or 27 percent of the 100 calories, within the guidelines for total fat. The guidelines are for the entire diet, not each individual food. Nevertheless, it is easy to see that a ½ cup serving of a premium ice cream that can contain 24 grams of fat (depending on brand and flavour) will have to be offset by a lot of steamed vegetables and other low-calorie foods. When reading labels it may also be helpful to visualise the fat. For example, 5 grams of fat is approximately equal to a pat of butter or margarine or a teaspoon of oil.

Preparing Food

Filling the cupboards and refrigerator with low-fat, low-cholesterol foods is half the battle. The other half is getting them on the table without adding too much fat, while still presenting them in a way the family will enjoy. The first step is to trim any excess fat before cooking. Even lean cuts of meat have exterior fat that can easily be trimmed. With poultry, removing the skin and the fat attached to it cuts the amount of fat to a little less than half. This is fine for most chicken and turkey dishes, but not practical for roasting, as the result is much too dry. In this case, removing the skin before eating is sufficient.

Fat that can't be eliminated before cooking can be drained off afterward. Be sure to drain the excess fat off ground meat before adding other ingredients. Refrigerate soups and stews until the fat congeals on top to make the removal of fat easy before reheating. If time does not allow preparing ahead, use a paper towel to soak up the surface oil or a skimmer to pour it off. A gravy separator (a pitcher with a spout that comes from the bottom) is another handy gadget for removing excess fat.

The second major way to control fat is via the method of cooking. Deep-frying, sautéing, breading and frying, basting with oil, and, with some exceptions, cooking in a casserole are methods that add unnecessary fat. Instead, try broiling, roasting, baking, steaming, poaching, and stir-frying.

With stirfrying, use only a little oil and be sure it is very hot, since foods soak up more oil when it is only warm. Blanching vegetables first cuts frying time and thus the amount of oil absorbed. Sauteing is acceptable if a nonstick pan or nonstick vegetable spray is used. Steaming isn't only for vegetables. Fish lends itself to steaming (a wok with a plate balanced above water level on overturned coffee-cups makes an inexpensive fish steamer). Chicken or fish can be baked in a foil package in the oven, which produces the same effect as steaming. For example, place pieces of chicken breast, seasoned with herbs and spices, in foil along with a selection of vegetables such as onions, mushrooms and zucchini, and perhaps a little white wine, and steam-bake for 20 minutes at 375° F.

Although many cooks like to experiment from time to time, most find that they develop a repertoire of 10 to 20 dishes that are repeated over and over. Perhaps the best way to ease gradually into low-fat eating is to work with family favourites and learn how to modify these recipes rather than to learn a whole new way of cooking. The three main principles of adapting recipes are reduction, substitution, and modification. Below are some tips for accomplishing this in some American stand-bys, but any recipe can be modified with a little analysis of its ingredients. Learn to determine why an ingredient is included in a recipe, and it will be easier to find an appropriate substitution or to decide whether it can be reduced or even eliminated. An ingredient may be there primarily to add flavour, or for bulk, texture, moisture, or eye appeal.

Crunchy nuts, for example, could be replaced by crunchy celery, jicama, or water chestnuts. Parmesan cheese topping on a casserole gives a flavourful browned appearance, but it can be mixed half and half with nonfat bread crumbs for the same effect. Or eliminate it entirely and mix the bread crumbs with herbs. Here are some other modifications and cooking tips:

— Steam vegetables with herbs, such as green beans with dill or carrots with basil, rather than adding butter.

— Use pureed potatoes or other vegetables instead of cream to thicken soups and stews. Barley, rice, and orzo can also add thickness to soup. In cream soups, use nonfat dry milk or evaporated skim milk instead of cream.

— Although its primary contribution is flavour, fat is sometimes added for moisture. In this case, fruit or vegetable juice, low-sodium broth, vinegar, wine, or beer can be used instead, depending on the dish.

— Tenderise lean meat by marinating it in herbs mixed with something acidic: tomato juice, citrus juice, vinegar, yogurt, or wine.

— Sugar can be reduced by at least one-third in baked goods without affecting the final product. Experiment to see if more can be removed.

— Dishes other than baked goods can be sweetened with undiluted frozen apple juice or pureed bananas or pears. Spices such as cinnamon add a sweet taste without calories. Although honey has a few more calories than sugar, it is much sweeter, so less is needed.

— Make mashed potatoes with skim milk or whip them with butter-substitute granules or some of the water in which they were cooked.

— Substitute ground turkey for ground beef and turkey or chicken cutlets for veal.

— Use 3 tablespoons of cocoa powder and 1 tablespoon of polyunsaturated oil in place of each ounce of baking chocolate.

— Try whipping very cold or partially frozen evaporated skim milk in place of heavy cream. (Chili the beaters and bowl in the freezer first.)

— Experiment with nonfat yogurt and light sour cream in place of traditional sour cream. In hot dishes, add 1 teaspoon of cornstarch for every cup of yogurt to keep it from separating on heating. Another sour cream substitute can be made from 1 cup of low-fat cottage cheese beaten with 1 tablespoon of lemon juice and 2 tablespoons of skim milk.

— Use one whole egg plus one white in place of two whole eggs, or use ¼ cup of egg substitute. In some recipes, two egg whites can substitute for a whole egg.

— Cut the amount of meat and increase the amount and variety of vegetables in stews.

— Salt can be eliminated from virtually any dish except yeast breads, in which it is needed to control rising. Since salt is an acquired taste, by gradually restricting it in the diet, it is possible to get used to tasting less and less of it, Herbs, spices, garlic, onions, citrus juices, and vinegars such as fruit or rice vinegars can be used instead to enhance the flavour of many foods.

— Instead of using a ham bone to add a smoky flavour to pea soup, add a sweet red pepper that has been roasted, slightly charred, and peeled.

For those looking for inspiration, there are a number of good low-fat cookbooks on the market. Ethnic cookbooks, particularly the dozen or so cuisines of the Mediterranean area, offer some excellent choices as well.

Fast-food Eating

For more than a decade, Americans have been eating one out of three meals away from home-in restaurants, school or company cafeterias, and fast food outlets. With more and more women joining the work force, this trend can only grow stronger. Those for whom dining out still means a special occasion generally can celebrate without concern about diet. Those who find themselves regularly relying on someone else's menus will need to devise a strategy to stay in control of what they eat.

The strategy starts with choosing a restaurant. It will be easier to find low-fat entrees in a seafood restaurant than in one that specialises in continental cuisine (read heavily sauced). Restaurateurs who pride themselves on offering "family-style service," "over-stuffed sandwiches," "he-man portions," or "all you can eat" are less likely to be concerned about nutritional balance. Generally speaking, restaurants that offer mostly freshly prepared dishes rather than mass-pre-cooked, flash-frozen portions can be more flexible. These restaurants are often recognisable by their limited menus. Even if their cuisine tends toward butter and sauces, they are usually willing to offer plain broiled fish, chicken, or lean steak with the sauce served on the side or not at ail. If the choice of restaurants is limited by budget, convenience, or dining companions, it is still possible to put together an acceptable meal.

Variety of Foods in Diet

Consume a variety of foods balanced by a moderate intake of each food. A variety of foods is best because no one food meets all your nutrient needs. Human milk comes close to meeting all of an infant's needs, except that it provides only limited amounts of iron, vitamin D, and fluoride. Cow's milk contains very little iron; neither form of milk provides dietary fiber. Meat provides protein but little calcium. Eggs have no vitamin C and provide little calcium because the calcium is mostly in the shell. Thus, you need variety in your diet because the required nutrients are scattered among many different foods.

Health professionals have recommended the same basic diet and health plan for the past 30 years: Watch how much you eat, focus on the major food groups, and stay physically active. Whole grains, fruits, and vegetables have always been among the foods emphasised for our diet for the past 30 years. It is disappointing, however, that according to a recent survey conducted by The American Dietetic Association, two of five people in the United States believe that following a healthful diet means giving up foods they enjoy. To the contrary, a healthful diet requires only some simple planning and doesn't have to mean deprivation and misery. Besides, eliminating favourite foods typically doesn't work for "dieters" in the long run. The best plan consists of learning the basics of a healthful diet-a variety and balance of foods from all food groups and moderate consumption of all foods. Let's now fine-tune this advice by focusing on variety, balance, moderation, nutrient density, and energy density.

Variety in your diet means choosing a number of different foods within any given food group rather than eating the "same old thing" day after day. Variety makes meals more interesting and helps ensure that a diet contains sufficient nutrients. For example, carrots may be your favourite vegetable; however, if you choose carrots every day as your only vegetable source, you may miss out on the vitamin folate. Other vegetables, such as broccoli and asparagus, are rich sources of this nutrient. This concept is true of all classes of foods: fruits, vegetables, grains, and so on. Different foods within each class vary somewhat in the nutrients they contain, but they generally provide similar types of nutrients.

An added bonus of variety in the diet, especially within the fruit and vegetable groups, is the inclusion of a rich supply of what scientists call phytochemicals. These substances are not absolutely required elements of the diet. Still, many of these substances provide significant health benefits. Considerable research attention is focused on various phytochemicals in reducing the risk for certain diseases such as cancer. Because current multivitamin and mineral supplements contain few or none of these beneficial substances, they generally are available only from food.

Numerous population studies show reduced cancer among people who regularly consume fruits and vegetables. This is true for cancer of the gastrointestinal (GI) tract, breast, lung, and bladder. Researchers surmise that some phytochemicals present in the fruits and vegetables block the cancer process. For now, realise that cancer develops over many years via a multistep process. If an agent such as a phytochemical can block any one of the steps in this process, the chances that cancer will ultimately appear in the body are reduced. Other phytochemicals have been linked to a reduced risk of cardiovascular disease. Could it be that, because humans evolved on a wide variety of plant-based foods, the body developed with a need for these phytochemicals, along with the various nutrients present, to maintain optimal health?

It will likely take many years for scientists to unravel the important effects of the myriad of phytochemicals in foods, and it is unlikely that all will ever be available in supplement form. For this reason, leading cardiovascular disease and cancer researchers suggest that a diet rich in fruits, vegetables, and whole grains is the most reliable way to obtain the potential benefits of phytochemicals.

Not Overconsuming Any One Food

One way to balance your diet as you consume a variety of foods is to select foods from the five major food groups every day:

— Milk, yogurt, and cheese
— Meat, poultry, fish, dry beans, eggs, and nuts
— Vegetables
— Fruit
— Bread, cereal, rice, and pasta

A lunch consisting of a bean burrito with tomatoes accompanied by a glass of milk and an apple covers all groups. Fats, oils, and sweets can also be added to your diet in moderation to increase its flavour and help deliver certain nutrients, such as vitamin E and essential fatty acids.

Moderating Portion Size

Although moderating portion size is a good practice, eating moderately also requires planning your entire day's diet so that you don't overconsume nutrient sources. For example, if you eat something relatively high in fat, sugar, or energy, such as a bacon cheeseburger with a regular soft drink at a fast-food (quick-service) restaurant, you should eat other foods that are less concentrated sources of the same nutrients, such as fruits and salad greens that same day. This helps balance one's diet. If you prefer whole milk to low-fat or nonfat milk, reduce the fat elsewhere in your meals.

Try low fat salad dressings, or use jam rather than butter or margarine on toast. Overall, strive to simply moderate-rather than eliminate-intake of some foods. Although there are no "good" or "bad" foods as such, many North Americans have diets overloaded with high-fat foods (e.g., whole milk, doughnuts, French fries, hot dogs), white bread and related refined wheat products, and sugared soft drinks. Such diets lack the foundations of a healthy food plan-variety, balance, and moderation-and pose substantial risks for nutrition-related diseases.

Table 2. Some Phytochemical Compounds Under Study

Phytochemical	*Food Sources*
Allyl sulfides/organosulfurs	Garlic, onions, leeks
Saponins	Garlic, onions, licorice, legumes
Phenolic acids	All plants
Protease inhibitors	Soybeans and all other plants
Carotenoids	Orange, red, yellow fruits and vegetables (egg yolks are a source as well)
Monoterpenes	Oranges, lemons, grapefruit
Capsaicin	Chili peppers
Lignans	Flaxseed, berries, whole grains
Triterpenoids	Citrus fruit, mushrooms
Indoles	Cruciferous vegetables (broccoli, cabbage, kale)
Isothiocyanates	Cruciferous vegetables, especially broccoli
Phytosterols	Soybeans, other legumes, cucumbers, other fruits and vegetables
Flavonoids	Citrus fruit, onions, apples, grapes, red wine, tea, chocolate
Isoflavones	Soybeans, other legumes
Catechins	Tea
Ellagic acid	Strawberries, raspberries, grapes, apples, bananas
Anthocyanosides	Red, blue, and purple plants (eggplant, blueberries)
Curacumin	Turmeric
Dithiolthiones	Carrots
Fructooligosaccharides	Onions, bananas, oranges

Nutrient Density

Nutrient density has gained acceptance in recent years for assessing the nutritional quality of an individual food. To determine the nutrient density of a food, simply compare its vitamin or mineral content with the amount of energy it provides. A food is said to be nutrient dense if it provides a large amount of a nutrient for a relatively small amount of calories. The higher a food's nutrient density, the better it is as a nutrient source. Comparing the nutrient density of different foods is an easy way to estimate their relative nutritional quality.

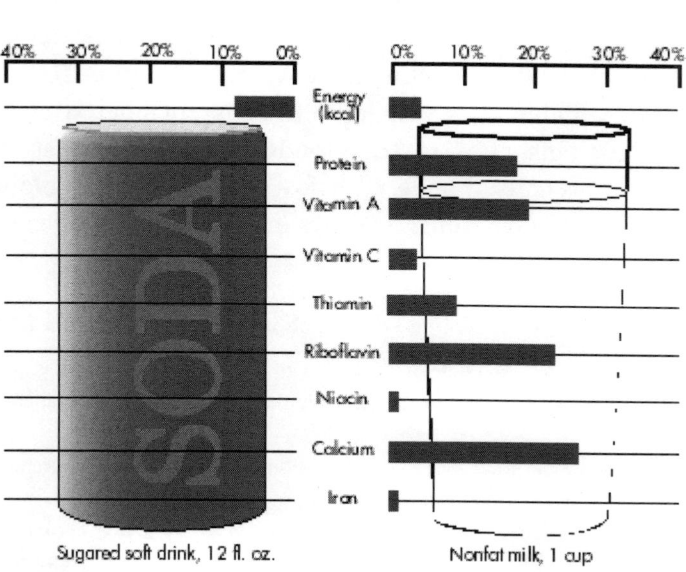

Figure 1. Comparison of the nutrient density of a sugared soft drink with that of nonfat milk

Generally, nutrient density is assessed with respect to individual nutrients. For example, many fruits and vegetables have a high content of vitamin C compared with their modest energy content: That is, they are nutrient-dense foods for vitamin C. Moreover, as Figure 1 shows, nonfat milk is much more nutrient dense than sugared soft drinks for many nutrients.

As noted previously, menu planning focuses mainly on the total diet-not on the selection of one critical food as key to an adequate diet. Nonetheless, nutrient-dense foods-such as nonfat and low-fat milk, lean meats, beans, oranges, carrots, broccoli, whole-wheat bread, and whole-grain breakfast cereals-do help balance less nutrientdense foods-such as cookies and potato chips-which many people like to eat. The latter are often called empty-calorie foods because they tend to supply much energy as sugar and/or fat but few other nutrients. Searching for nutrient-dense foods is especially important in some cases. For example, this strategy can aid diet planning for people who tend to consume little food energy, including some older people and those following weight-loss diets.

Energy (kcal) Density

Energy density is a concept that has captured the attention of nutrition scientists in recent years. Energy density of a food is determined by comparing energy (kcal) content with the weight of food. A food that is rich in calories but weights relatively very little is considered energy dense. Examples include nuts, cookies, fried foods in general, and fat-free snacks, such as fat-free pretzels. Foods with low energy density include fruits, vegetables, and any food that incorporates lots of water during cooking, such as oatmeal (Table 4).

Table 3. Tips for Including Foods Rich in Phytochemicals in a Diet

- — Include vegetables in main and side dishes. Add these to rice, omelets, potato salad, tuna salad, and pastas. Try broccoli or cauliflower florets, mushrooms, peas, carrots, corn, or peppers.
- — Look for quick-fixing grain side dishes in the supermarket. Pilafs, couscous, rice mixes, and tabbouleh are just a few that you'll find.
- — Choose fruit-filled cookies, such as fig bars, instead of sugar-rich cookies. Use fresh or canned fruit as a topping for puddings, hot or cold cereal, pancakes, and frozen desserts.
- — Put raisins, grapes, apple chunks, pineapples, grated carrots, zucchini, or cucumber into coleslaw, chicken salad, or tuna salad.
- — Be creative at the salad bar: Try fresh spinach, leaf lettuce, red cabbage, zucchini, yellow squash, cauliflower, peas, mushrooms, or red or yellow peppers.
- — Pack fresh or dried fruit for snacks away from home instead of grabbing a candy bar or going hungry.
- — Add slices of cucumber, zucchini, spinach, or carrot slivers to the lettuce and tomato on your sandwiches.
- — Try one or two vegetarian meals per week, such as beans and rice or pasta; Chinese vegetable stir fry; or spaghetti, squash, and tomato sauce.
- — When daily protein intake more than meets recommended amounts, reduce the meat, fish, or poultry in casseroles, stews, and soups by one-third to one-half and add more vegetables and legumes.
- — Keep a bowl of fresh vegetables in the refrigerator for snacks.
- — Choose fruit or vegetable juices instead of soft drinks, and preferably 100% varieties.
- — Substitute tea for coffee or soft drinks on a regular basis.

— Have a bowl of fruit on hand.
— Switch from crisphead lettuce to leaf lettuce, such as romaine.
— Use salsa as a dip for chips in place of creamy dips.
— Choose whole-grain breakfast cereals, breads, and crackers.
— Flavour food with plenty of herbs and spices, including ginger, rosemary, basil, thyme, garlic, parsley, and chives in place of salt.
— Experiment with soy products, such as tofu, soy milk, soy protein isolate, and roasted soybeans.

Table 4. Energy Density of Common Foods (Listed in Relative Order)

Very Low Energy Density (0.6 kcal per gram)	*Low Energy Density (0.6 to 1.5 kcal per gram)*
Lettuce	Whole milk
Tomatoes	Oatmeal
Strawberries	Cottage cheese
Broccoli	Beans
Salsa	Bananas
Grapefruit	Broiled fish
Nonfat milk	Fat-free yogurt
Carrots	Ready-to-eat breakfast cereals with 1% low-fat milk
Vegetable soup	Plain baked potato
	Cooked rice
	Spaghetti noodles
Medium Energy Density (1.5 to 4 kcal per gram)	*High Energy Density (-4 kcal per gram)*
Eggs	Graham crackers
Ham	Fat-free sandwich cookies
Pumpkin pie	Chocolate
Whole-wheat bread	Chocolate chip cookies
Bagels	Tortilla chips
White bread	Bacon
Raisins	Potato chips
Cream cheese	Peanuts
Cake with frosting	Peanut butter
Pretzels	Mayonnaise
Rice cakes	Butter or margarine
	Vegetable oils

Researchers have shown that having low-energy-density foods in a meal contributes to satiety without contributing many calories. This is because we probably consume a constant weight of food at a meal, rather than a constant number of calories. How this constant weight of food is regulated is not known, but careful laboratory studies show that people consume fewer calories in a meal if the food choices tend to be low in energy density, compared with foods high in energy density. Following (or maintaining) such a diet low in energy density can aid in losing weight.

Overall, foods with lots of water and dietary fiber provide a low-energy-density contribution to a meal and help one feel full, whereas foods with high energy density -especially those high in fat-must be eaten in greater amounts in order to contribute to fullness. This is one more reason to support a diet rich in fruits, vegetables, and whole grains, a pattern that also is typical of many ethnic diets throughout the world. Still, favourite foods, even if they are high in energy density, can have a place in your dietary pattern, but you will have to plan for them. For example, chocolate is a very energy-dense food, but a small portion at the end of a meal can supply a satisfying finale. In addition, foods with high energy density can help people with poor appetites, such as older people, to maintain or gain weight.

NUTRITIONAL TOOLS AND GUIDELINES FOR PLANNING HEALTHY DIETS

Nutritional Health of Body

The body's nutritional health is determined by the sum of its nutritional state with respect to each needed nutrient. Three general categories are recognised: desirable nutrition, undernutrition, and overnutrition. Maintaining a state of desirable nutrition is the basis for establishing human nutrient needs. The common term malnutrition can refer to either overnutrition or undernutrition. Neither state is conducive to good health.

Desirable nutrition

The nutritional state for a particular nutrient is desirable when body tissues have enough of the nutrient to support normal metabolic functions as well as surplus stores that can be used in times of increased need. A desirable nutritional state can be achieved by obtaining essential nutrients from a variety of foods.

Undernutrition

Undernutrition occurs when nutrient intake does not meet nutrient needs.

Stores are then used up and health declines. Many nutrients are in high demand due to the constant state of cell loss and later regeneration in the body, such as in the gastrointestinal tract. For this reason, certain nutrient stores are exhausted rapidly, such as for many of the B vitamins. In turn, a regular intake is needed. In addition, some women in North America do not consume sufficient iron to meet monthly losses and eventually deplete their iron stores. Once availability of a nutrient falls sufficiently low, biochemical evidence, in which the body's metabolic processes slow or stop, appears. At this state of deficiency there are not outward symptoms, thus it is termed a subclinical deficiency. A subclinical deficiency can go on for some time before clinicians are able to detect its effects.

Table 5. Conducting an Evaluation of Nutritional Health

Component	*Example*
Background histories	Medical history, including current diseases and past surgeries
	Medications history
	Social history (marital status, cooking facilities)
	Family history
	Economic status
	Education attainment
Nutritional parameters	Anthropometric assessment: height, weight, skinfold thickness, arm muscle circumference, and other parameters
	Biochemical (laboratory) assessment of blood and urine: enzyme activities, concentrations of nutrients or their by-products
	Clinical assessment (physical examination): general appearance of skin, eyes, and tongue; rapid hair loss; sense of touch; ability to walk
	Diet history: usual intake or record or previous days' meals

Eventually clinical symptoms will develop, sometimes taking many years, and may result in clinical evidence of a deficiency; perhaps in the skin, hair, nails, tongue, or eyes. Often, clinicians do not detect a problem until a deficiency produces such results, such as in a vitamin C deficiency.

Overnutrition

Prolonged consumption of more nutrients than the body needs can lead to overnutrition. In the short run, for instance a week or two, overnutrition may cause only a few symptoms, such as stomach distress from excess dietary fiber or iron intake. But if kept up, some nutrients may increase to toxic amounts, which can lead to serious disease. For example, too much vitamin A can have negative effects, particularly in children and pregnant women. The most common type of overnutrition in industrialised nations-excess intake of energy-yielding nutrients-often leads to obesity. In the long run, obesity can then lead to other serious diseases, such as certain forms of diabetes and cancer. Use the website shapeup.org to learn more about this problem.

For most vitamins and minerals, the gap between desirable intake and overnutrition is wide. Therefore, even if people take a typical multivitamin and mineral supplement daily, they probably won't receive a harmful amount of any nutrient. The gap between optimal intake and overnutrition is narrowest for vitamin A, calcium, iron, copper, and other minerals. Thus, if you take nutrient supplements, keep a close eye on your total vitamin and mineral intake both from food and from supplements to avoid toxicity.

Measurement of Nutritional State

To find out how nutritionally fit you are, a nutritional assessment-either whole or in part-needs to be performed. Generally, this is performed by a physician and often with the aid of a registered dietitian.

Evaluating the ABCDEs

Five components in combination further add to the complete nutritional picture. Anthropometric measurements of height, weight, body skinfolds, and body circumferences are an excellent first line of attack. They are easy to obtain and are generally reliable. However, an in-depth examination of nutritional health is impossible without the more expensive process of biochemical assessment. This involves the measurement of specific blood enzyme activities and of the concentrations of nutrients and nutrient by-products in the blood, urine, and feces.

A clinical examination would follow, during which a health professional would search for any physical evidence of diet-related diseases. Then, a diet history, documenting at least the previous few days' intake, would look into possible problem areas. Finally, current economic status is added to the picture, such as the ability to purchase and prepare appropriate foods needed to maintain health. Now the true nutritional state of a person emerges. Together these activities form the ABCDEs of nutritional assessment: anthropometric measurement, biochemical assessment, clinical examination, diet history, and economic status.

Limitations of Nutritional Assessment

As mentioned, a long time may elapse between the initial development of poor nutritional health and the first clinical evidence of a problem. Recall that a diet high in saturated (typically solid) fat often increases blood cholesterol concentration, but without producing any clinical evidence for years. However, when the blood vessels become sufficiently blocked by cholesterol and other materials, chest pain during physical activity or a heart attack may occur. Much current nutrition research aims to develop better methods for early detection of nutrition-related problems such as this.

Another example in the delay of evidence that serious consequences are occurring is with a calcium deficiency, a particularly relevant issue for adolescent females. Many young women consume well below the needed amount of calcium but often suffer no ill effects in their younger years. However, women whose bone structures do not reach full potential during the years of growth are likely to face an increased risk for osteoporosis later in life.

Furthermore, clinical evidence of nutritional deficiencies is often not very specific, such as diarrhea, an irregular walk, and facial sores. These may have different causes. Long lag times and vague evidence often make it difficult to establish a link between an individual's current diet and nutritional state.

Guidelines for Food Choice

There are various guidelines for planning healthy diets.

Food guide pyramid-a menu-planning tool

Since the early twentieth century, researchers have worked to clarify the science of nutrition into practical terms, so that people with no special training could estimate whether their nutritional needs were being met. A seven-food-group plan, based on foods traditionally eaten by people in the United States, was one of the first formats. Daily food choices had to include items from each group. This plan had been simplified by the mid-1950s to a four-food-group plan: a milk group, a meat group, a fruit and vegetable group, and a bread and cereal group.

The entire plan was designed to provide a minimum foundation for a diet, and represented about 1200 to 1400 kcal per day. Other food choices were to be added to meet daily energy needs. Today, the Food Guide

Pyramid, which is designed to represent a total diet providing sufficient protein, vitamins, and minerals, is widely advocated for diet planning (Fig. 2.). This pyramid goes beyond earlier food guides to suggest a pattern of food choices for the entire day, rather than simply a foundation diet.

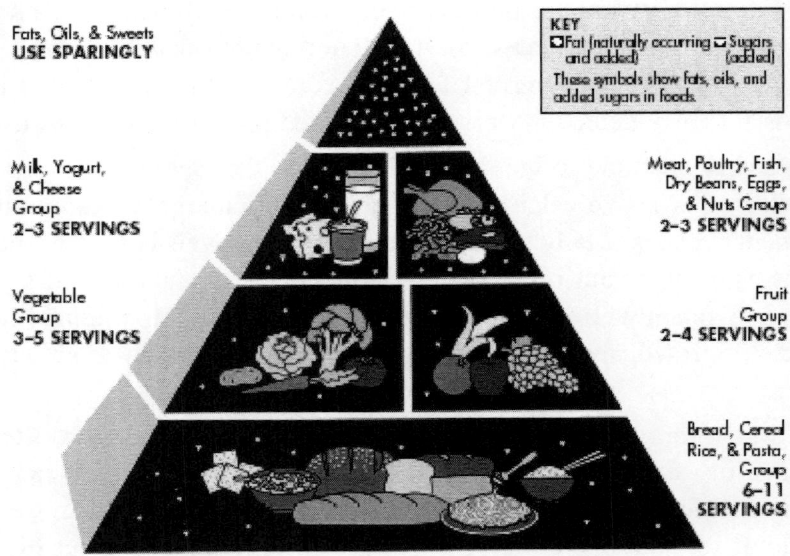

Figure 2. USDA's Food Guide Pyramid.

Components of the food guide pyramid

The number of servings to consume from each food group in the current Food Guide Pyramid depends on a person's age and energy needs. Serving size is also adjusted downward for young children. The plan for an adult over age 18 essentially consists of the following:

— 2 servings from the milk, yogurt, and cheese group
— 2 to 3 servings from the meat, poultry, fish, dry beans, eggs, and nuts group (5 to 7 ounces total)
— 3 to 5 servings from the vegetable group
— 2 to 4 servings from the fruit group
— 6 to 11 servings from the bread, cereals, rice, and pasta group

Note also that some food choices will contain servings of more than one food group (e.g., lasagna contains pasta, cheese, and tomatoes, and likely ingredients from other food groups as well). For some population groups-children, and teenagers; and teenagers who are pregnant or breastfeeding-three servings of the milk, yogurt, and cheese group are recommended due to higher calcium needs. The same is also true for older adults (51 years). Alternately, some of those servings also could be replaced with calciumfortified foods or a calcium supplement.

Foods in a final category at the tip of the pyramid, which is not a group per se, include fats, oils, and sweets. These can be eaten to help meet individual energy needs but should not replace foods from other groups.

Menu Planning with the Food Guide Pyramid

Table 6 illustrates a 1-day menu based on the Food Guide Pyramid. Remember the following points when using the Food Guide Pyramid to plan daily menus:

1. The guide does not apply to infants or children under 2 years of age.
2. No one food is absolutely essential to good nutrition. Each food is deficient in at least one essential nutrient.
3. No one food group provides all essential nutrients in adequate amounts. Each food group makes an important, distinctive contribution to nutritional intake.
4. Variety is the key to success of the guide and is first guaranteed by choosing foods from all the groups. Furthermore, one should consume a variety of foods within each group, except possibly in the milk, yogurt, and cheese group.
5. The foods within a group may vary widely with respect to nutrient and energy content. For example, the energy content of 3 ounces of baked potato is 98 kcal, whereas that of 3 ounces of potato chips is 470 kcal.

Overall, the Food Guide Pyramid incorporates the foundations of a healthy diet: variety, balance, and moderation. The nutritional adequacy of diets planned using this tool, however, depends on selection of a variety of foods. In addition, to ensure enough vitamin E, vitamin B-6, magnesium, and zinc-nutrients sometimes low in diets based on this plan-consider the following advice:

1. Choose primarily low-fat and nonfat items from the milk, yogurt, and cheese group. By reducing energy intake in this way, you can select more items from other food groups. If milk causes intestinal gas and bloating, emphasise yogurt and cheese.
2. Include plant foods that are good sources of proteins, such as beans and nuts, at least several times a week because many are rich in vitamins (such as vitamin E), minerals (such as magnesium), and dietary fiber.
3. For vegetables and fruits, try to include a dark green vegetable for vitamin A and a vitamin C-rich fruit, such as an orange, every day. Don't focus primarily on potatoes for your vegetable choices. Surveys show that only 25% of adults eat a green vegetable on any given day. Increased consumption of these foods is important because they contribute vitamins, minerals, dietary fiber, and phytochemicals.
4. Choose whole-grain varieties of breads, cereals, rice, and pasta often because they contribute vitamin E and dietary fiber. A plate about two-thirds covered by grains, fruits, and vegetables and one-third or less covered by protein-rich foods promotes this diet advice. As well, a daily serving of a whole-grain, ready-to-eat breakfast cereal is an excellent choice because the vitamins (such as vitamin B-6) and minerals (such as zinc) typically added to it, along with dietary fiber, help fill in the potential gaps.
5. Include some plant oils on a daily basis, such as those in salad dressing, and eat fish at least twice a week. This supplies you with health-promoting fatty acids.

Following the Food Guide Pyramid makes it possible to create daily diets containing as few as 1600 to 1800 kcal (Table 6), sufficient for a sedentary adult or an older person. Not following this advice can leave a diet 1600 to 1800 kcal short on the nutrients just mentioned. Recall that excessive consumption of any one food-even those considered "healthy"-is also undesirable and possibly risky.

If 1600 to 1800 kcal represents too much food energy for you, you should first consider becoming more physically active rather than eating less. Obtaining enough nutrients from a diet that supplies fewer than 1600 kcal per day is very difficult. If you can't increase your energy output, you can make a special attempt to choose some nutrient-fortified foods regularly (e.g., ready-to-eat breakfast cereals) or take a balanced multivitamin and mineral supplement.

Table 6. Putting the Food Guide Pyramid into Practice

Meal	*Servings/Food Group*
Breakfast	
1 small, peeled orange	1 fruit
¾ cup Healthy Choice Low-fat Granola	1 bread
with ½ cup nonfat milk	½ milk
½ toasted, small raisin bagel	1 bread
with 1 tsp soft margarine	1 fat/sweet
Optional: coffee or tea	
Lunch	
Ham sandwich	
2 slices whole-wheat bread	2 bread
2 oz ham	1 meat
2 tsp	mustard
1 small apple	1 fruit
2 oatmeal-raisin cookies (small)	2 fat/sweet
Optional: diet soft drink	
3 PM Study Break	
6 whole wheat crackers	2 bread
1 tbsp peanut butter	¼ meat
½ cup nonfat milk	½ milk
Dinner	
Lettuce salad	
1 cup romaine lettuce	1 vegetable
½ cup sliced tomatoes	1 vegetable
1 tbsp Italian dressing	11/2 fat/sweet
½ grated carrot	1 vegetable
3 oz broiled salmon	1 meat
½ cup rice	1 bread
½ cup green beans	1 vegetable
with 1 tsp margarine	1 fat/sweet
Optional: coffee or tea	
Late-night snack	
1 cup "light" fruit yogurt	1 milk
Nutrient breakdown	
	1800 kcal
Carbohydrate	56% of kcal
Protein	18% of kcal
Fat	26% of kcal

The average North American diet, based on surveys, fails to meet the serving recommendations in the Food Guide Pyramid for many food groups. For example, the average diet included only 1 to 2 fruit servings (rather than the recommended 2 to 4 servings) and only 2 to 3 vegetable servings (rather than 3 to 5 servings), and much of that comes from potatoes, not a particularly nutrient-dense vegetable choice. Overall, fruits and vegetables are the most underrepresented groups. In contrast, the fats, oils, and sweets are well represented.

Current Diet Rate

Regularly comparing your daily food intake with the Food Guide Pyramid recommendations is a relatively simple way to evaluate your overall diet. Strive to meet the recommendations. If that is not possible, identify the nutrients that are low in your diet based on the nutrients found in each food group. For example, if you do not consume enough servings from the milk, yogurt, and cheese group, your calcium intake is most likely too low. Armed with this knowledge, find foods that you enjoy that supply those nutrients, such as calcium-fortified orange juice. Customising the Food Guide Pyramid to accommodate your own food habits may seem a daunting task now, but it is not difficult once you gain some additional nutrition knowledge.

Dietary guidelines-another tool for menu planning

The Food Guide Pyramid was designed to help meet nutritional needs for carbohydrate, protein, fat, vitamins, and minerals. However, most of the major chronic "killer" diseases in North America, such as cardiovascular disease, cancer, and alcoholism, are not primarily associated with deficiencies of these nutrients. Deficiency diseases such as scurvy (vitamin C deficiency) and pellagra (niacin deficiency) are no longer common. For many North Americans, the primary dietary culprit is an overconsumption of one or more of the following: energy, saturated fat, cholesterol, alcohol, and sodium (salt).

Underconsumption of calcium, iron, folate and other B-vitamins, vitamin D, vitamin E, zinc, or dietary fiber is also a problem for some people, but easy to fix as the major dietary problems are addressed. In response to concerns regarding these killer disease patterns in the United States, since 1980 the USDA and Department of Health and Human Services (DHHS) have published Dietary Guidelines to aid diet planning.

Use of the Dietary Guidelines

The Dietary Guidelines are designed to promote adequate vitamin and mineral intake. The guidelines also emphasise changes that will reduce the risk of obesity, hypertension, cardiovascular disease, type 2 diabetes, alcoholism, and food-borne illness.

Table 7. Advice for Applying the Dietary Guidelines to Practical Situations

If You Usually Eat This	Eat This More Often:
White bread	Whole-wheat bread (fewer nutrients lost in refinement/processing and more dietary fiber)
Sugared breakfast cereal	Low-sugar (and high-fiber) cereal (use the calories you save for a side dish of fruit)
Cheeseburger and French fries	Hamburger (hold the mayonnaise) and baked beans (for less fat and cholesterol, and the benefits of plant proteins)
Potato salad at the salad bar	Three-bean salad
Doughnut, chips, salty snack foods	Bran muffin or bagel (little or no cream cheese)
Soft drinks	Diet soft drinks (save the kcal for more nutritious foods)
Boiled vegetables	Steamed vegetables (for more nutrient retention)
Canned vegetables	Frozen vegetables (fewer nutrients lost in processing)
Fried meats	Broiled meats (watch the fat drain away)
Fatty meats, such as ribs	Lean meats, such as ground round (also, eat chicken and fish often)
Whole milk and ice cream	Low-fat or nonfat milk and sherbet or frozen yogurt (to reduce saturated fat intake)
Mayonnaise or sour cream salad dressing	Oil and vinegar dressings or diet varieties (to save calories)
Cookies for a snack	Popcorn (air popped with minimal margarine or butter)
Heavily salted foods	Foods flavored primarily with herbs, spices, lemon juice

The Dietary Guidelines are not difficult to implement. In addition, this overall diet approach is not especially expensive, as some people suspect. Fruits, vegetables, and low-fat and nonfat milk are no more expensive than the chips, cookies, and sugared soft drinks they should in part replace. Note also that diet recommendations for adults have been issued by other scientific groups, such as the American Heart Association, U.S. Surgeon General,

National Academy of Sciences, American Cancer Society, Canadian Ministries of Health, and World Health Organisation. All are consistent with the spirit of the Dietary Guidelines. These groups encourage people to modify their eating behaviour in ways that are both healthful and pleasurable.

When using the Dietary Guidelines, you should consider your own state of health. Make specific changes and see whether they are effective. Note that results are sometimes disappointing, even when you are following a diet change very closely. Some people can eat a lot of saturated fat and still keep blood cholesterol under control. Other people, unfortunately, have high blood cholesterol even if they eat a diet low in saturated fat. Thus, we have individual nutritional needs and risks of developing certain diseases. One's diet should be planned with this in mind, responding to one's current health status and family history for specific diseases. However, tailoring a unique nutrition programme for every North American citizen is unrealistic. The Food Guide Pyramid and the Dietary Guidelines provide typical adults with simple advice, which can be actively practiced by anyone willing to take a step toward good health.

Today, nearly all foods sold in the grocery store must be labeled with the product name, name and address of the manufacturer, amount of product in the package, and ingredients listed in descending order by weight. This food and beverage labelling is monitored in the United States by government agencies such as Food and Drug Administration (FDA). The listing of certain food constituents also is required-specifically, on a Nutrition Facts panel. Use this information to learn more about what you eat. The following components must be listed: total kcal, kcal from fat, total fat, saturated fat, cholesterol, sodium, total carbohydrate, dietary fiber, sugars, protein, vitamin A, vitamin C, calcium, and iron. In addition to these required components, manufacturers can choose to list polyunsaturated and monounsaturated fat, potassium, dietary fiber, and others. Listing these components is required, however, if a claim is made about the health benefits of the specific nutrient or if the food is fortified with that nutrient.

The percentage of the Daily Value (% Daily Value) is usually given for each nutrient per serving. It is important to understand that these percentages are based on a 2000 kcal diet. In other words, they are not as applicable to people who require considerably more or less than 2000 kcal per day with respect to fat and carbohydrate intake. Serving sizes on the Nutrition Facts panel must be consistent among similar foods. This means that all brands of ice cream, for example, must use the same serving size on their label.

Many manufacturers list the Daily Values set for dietary components such as fat, cholesterol, and carbohydrate on the Nutrition Facts panel. This can be useful as a reference point. As noted before, they are based on 2000 kcal; if the label is large enough, amounts based on 2500 kcal are listed as well for total fat, saturated fat, carbohydrate, and dietary fiber.

Food labelling

Foods such as fresh fruits and vegetables, fish, meats, and poultry currently are not required to have Nutrition Facts labels. However, many grocers and some meat packers have voluntarily chosen to provide their customers with information on these products. Nutrition Facts labels on meat products will also likely be required in the coming years. The next time you are at the grocery store, ask where you might find information on the fresh products that do not have a Nutrition Facts panel. You will likely find a poster or pamphlet near the product; often, these pamphlets contain recipes that use your favourite fruit, vegetable, or cut of meat. They may even assist you in your endeavour to improve your diet.

Because protein deficiency is not a public health concern in the United States, declaration of the % Daily Value for protein is not mandatory on foods for people over 4 years of age. If the % Daily Value is given on a label, FDA requires that the product be analysed for protein quality. Because this procedure is expensive and time-consuming, many companies opt not to list a % Daily Value for protein rather than undergo the expense. However, labels on food for infants and children under 4 years of age must include the % Daily Value for protein, as must the labels on any food carrying a claim about protein content.

As a marketing tool directed toward the health-conscious consumer, food manufacturers are asserting that their products have all sorts of health benefits. This campaign began in earnest in 1984, when the Kellogg Company, in conjunction with The National Cancer Institute, printed a health claim on its "high-fiber" cereals, stating that dietary fiber may help prevent certain forms of cancer. This type of label message was not allowed at the time and caused a heated debate among nutrition scientists. After reviewing hundreds of comments on the proposed rule allowing health claims, FDA, which has legal oversight over most food products, decided to permit this and other health claims with certain restrictions.

Currently, FDA limits the use of health messages to specific diseases in which there is significant scientific agreement concerning the relationship between a nutrient, food, or food constituent and the disease. The claims allowed at this time may show a link between the following:

— A diet with enough calcium and a reduced risk of osteoporosis
— A diet low in total fat and a reduced risk of some cancers
— A diet low in saturated fat and cholesterol and a reduced risk of cardiovascular disease (typically referred to as heart disease on the label)
— A diet rich in dietary fiber-containing grain products, fruits, and vegetables and a reduced risk of some cancers
— A diet low in sodium and high in potassium and a reduced risk of hypertension and stroke
— A diet rich in fruits and vegetables and a reduced risk of some cancers
— A diet adequate in the synthetic form of the vitamin folate (called folic acid) and a reduced risk of neural tube defects (a type of birth defect)
— Use of sugarless gum and a reduced risk of tooth decay, especially when compared with foods high in sugars and starches
— A diet rich in fruits, vegetables, and grain products that contain fiber and a reduced risk of cardiovascular disease. Oats (oatmeal, oat bran, and oat flour) and psyllium are two fiber-rich ingredients that can be singled out in reducing the risk of cardiovascular disease, as long as the statement also says the diet should also be low in saturated fat and cholesterol
— A diet rich in whole-grain foods and other plant foods, as well as low in total fat, saturated fat, and cholesterol, and a reduced risk of cardiovascular disease and certain cancers
— A diet low in saturated fat and cholesterol that also includes 25 g of soy protein and a reduced risk of cardiovascular disease. The statement "one serving of the (name food) provides g of soy protein" must also appear as part of the health claim
— A diet rich in potassium and a reduced risk of stroke

— Omega-3 fatty acids from oils present in fish and a reduced risk of cardiovascular disease

— Margarines containing plant stanol and sterol esters and a reduced risk of cardiovascular disease

In addition, before a health claim can be made for a food product, it must meet two general requirements. First, the food must be a "good source" (before fortification) of dietary fiber, protein, vitamin A, vitamin C, calcium, or iron. Second, a single serving of the food product cannot contain more than 13 grams of fat, 4 grams of saturated fat, 60 milligrams of cholesterol, or 480 milligrams of sodium. If a food exceeds any one of these amounts, no health claim can be made for it, despite its other nutritional qualities. For example, even though whole milk is high in calcium, its label can't make the health claim about calcium and osteoporosis because whole milk contains 5 grams of saturated fat per serving.

References

Barker, Helen M. (2002). *Nutrition and dietetics for health care*. Edinburgh: Churchill Livingstone. p. 17.

Berg J, Tymoczko JL, Stryer L (2002). *Biochemistry* (5th ed.). San Francisco: W.H. Freeman. p. 603.

Commission on Life Sciences. (1985). *Nutrition Education in US Medical Schools*, p. 4. National Academies Press.

Shils et al. (2005). *Modern Nutrition in Health and Disease*. Lippincott Williams and Wilkins.

4

Carbohydrates and Nutrition

A carbohydrate is an organic compound that consists only of carbon, hydrogen, and oxygen, usually with a hydrogen:oxygen atom ratio of 2:1 (as in water); in other words, with the empirical formula $Cm(H_2O)_n$. Carbohydrates are not technically hydrates of carbon. Structurally it is more accurate to view them as polyhydroxy aldehydes and ketones.

The carbohydrates (saccharides) are divided into four chemical groupings: monosaccharides, disaccharides, oligosaccharides, and polysaccharides. In general, the monosaccharides and disaccharides, which are smaller (lower molecular weight) carbohydrates, are commonly referred to as sugars. The word saccharide comes from the Greek word s???a??? (sákkharon), meaning "sugar." While the scientific nomenclature of carbohydrates is complex, the names of the monosaccharides and disaccharides very often end in the suffix -ose. For example, blood sugar is the monosaccharide glucose, table sugar is the disaccharide sucrose, and milk sugar is the disaccharide lactose (see illustration).

Carbohydrates perform numerous roles in living organisms. Polysaccharides serve for the storage of energy (e.g., starch and glycogen), and as structural components (e.g., cellulose in plants and chitin in arthropods). The 5-carbon monosaccharide ribose is an important component of coenzymes (e.g., ATP, FAD, and NAD) and the backbone of the genetic molecule known as RNA. The related deoxyribose is a component of DNA. Saccharides and their derivatives include many other important biomolecules that play key roles in the immune system, fertilization, preventing pathogenesis, blood clotting, and development.

In food science and in many informal contexts, the term carbohydrate often means any food that is particularly rich in the complex carbohydrate starch (such as cereals, bread, and pasta) or simple carbohydrates, such as sugar (found in candy, jams, and desserts).

Foods high in carbohydrate include fruits, sweets, soft drinks, breads, pastas, beans, potatoes, bran, rice, and cereals. Carbohydrates are a common source of energy in living organisms; however, no carbohydrate is an essential nutrient in humans.

Carbohydrates are not necessary building blocks of other molecules, and the body can obtain all its energy from protein and fats. The brain and neurons generally cannot burn fat for energy, but use glucose or ketones. Humans can synthesize some glucose (in a set of processes known as gluconeogenesis) from specific amino acids, from the glycerol backbone in triglycerides and in some cases from fatty acids. Carbohydrate and protein contain 4 calories per gram, while fats contain 9 calories per gram. In the case of protein, this is somewhat misleading as only some amino acids are usable for fuel.

Organisms typically cannot metabolize all types of carbohydrate to yield energy. Glucose is a nearly universal and accessible source of calories. Many organisms also have the ability to metabolize other monosaccharides and Disaccharides, though glucose is preferred. In Escherichia coli, for example, the lac operon will express enzymes for the digestion of lactose when it is present, but if both lactose and glucose are present the lac operon is repressed, resulting in the glucose being used first (see: Diauxie). Polysaccharides are also common sources of energy. Many organisms can easily break down starches into glucose, however, most organisms cannot metabolize cellulose or other polysaccharides like chitin and arabinoxylans. These carbohydrates types can be metabolized by some bacteria and protists. Ruminants and termites, for example, use microorganisms to process cellulose. Even though these complex carbohydrates are not very digestible, they may comprise important dietary elements for humans. Called dietary fiber, these carbohydrates enhance digestion among other benefits.

Based on the effects on risk of heart disease and obesity, the Institute of Medicine recommends that American and Canadian adults get between 45–65% of dietary energy from carbohydrates. The Food and Agriculture Organization and World Health Organization jointly recommend that national dietary guidelines set a goal of 55–75% of total energy from carbohydrates, but only 10% directly from sugars (their term for simple carbohydrates).

Classification

Historically nutritionists have classified carbohydrates as either simple or complex. However, the exact delineation of these categories is ambiguous. Today, simple carbohydrate typically refers to monosaccharides and disaccharides and complex carbohydrate means polysaccharides (and oligosaccharides). However, the term complex carbohydrate was first used in slightly different context in the U.S. Senate Select Committee on Nutrition and Human Needs publication Dietary Goals for the United States (1977). In this work, complex carbohydrate were defined as "fruit, vegetables and whole-grains". Some nutritionists use complex carbohydrate to refer to any sort of digestible saccharide present in a whole food, where fiber, vitamins and minerals are also found (as opposed to processed carbohydrates, which provide calories but few other nutrients).

Some simple carbohydrates(e.g. fructose) are digested very slowly, while some complex carbohydrates (starches), especially if processed, raise blood sugar rapidly. The speed of digestion is determined by a variety of factors including which other nutrients are consumed with the carbohydrate, how the food is prepared, individual differences in metabolism, and the chemistry of the carbohydrate.

The USDA's Dietary Guidelines for Americans 2010 call for moderate- to high-carbohydrate consumption from a balanced diet that includes six one-ounce servings of grain foods each day, at least half from whole grain sources and the rest from enriched.

The glycemic index (GI) and glycemic load concepts have been developed to characterize food behavior during human digestion. They rank carbohydrate-rich foods based on the rapidity and magnitude of their effect on blood glucose levels. Glycemic index is a measure of how quickly food glucose is absorbed, while glycemic load is a measure of the total absorbable glucose in foods. The insulin index is a similar, more recent classification method that ranks foods based on their effects on blood insulin levels, which are caused by glucose (or starch) and some amino acids in food.

Carbohydrates in Nutrition

Carbohydrates are polyhydroxy aldehydes, ketones, alcohols, acids, their simple derivatives and their polymers having linkages of the acetal type. They may be classified according to their degree of polymerization and may be divided initially into three principal groups, namely sugars, oligosaccharides and polysaccharides. Each of these three groups may be subdivided on the basis of the monosaccharide composition of the individual carbohydrates. Sugars comprise monosaccharides, disaccharides and polyols (sugar alcohols); oligosaccharides include malto-oligosaccharides, principally those occurring from the hydrolysis of starch, and other oligosaccharides, e.g. a -galactosides (raffinose, stachyose etc.) and fructo-oligosaccharides; the final group are the polysaccharides which may be divided into starch (a -glucans) and non-starch polysaccharides of which the major components are the polysaccharides of the plant cell wall such as cellulose, hemicellulose and pectin.

Total Carbohydrate

Although the individual components of dietary carbohydrate are readily identifiable, there is some confusion as to what comprises total carbohydrate as reported in food tables. Two principal measures of total carbohydrate are used, firstly, that derived by "difference" and secondly the direct measurement of the individual components which are then combined to give a total. Calculating carbohydrates by "difference" has been used since the turn of the century. The protein, fat, ash and moisture content of a food are determined, subtracted from the total weight of the food and the remainder, or "difference", is considered to be carbohydrate.

There are, however, a number of problems with this approach to total carbohydrate analysis in that the "by difference" figure includes a number of non-carbohydrate components such as lignin, organic acids, tannins, waxes, and some Maillard products. In addition to this error, it combines all of the analytical errors from the other analyses. Finally, a single global figure for carbohydrates in food is uninformative because it fails to identify the many types of carbohydrates in a food and thus to allow some understanding of the potential physiological properties of those carbohydrates.

In deciding how to classify dietary carbohydrate the principal problem is to reconcile the various chemical divisions of carbohydrate with that which reflects physiology and health. A classification based purely on chemistry does not allow a ready translation into nutritional terms since each of the major classes of carbohydrate have a variety of physiological effects. However, a classification based on physiological properties also creates a number of problems in that it requires a single effect to be considered as over-ridingly important and to be used as the basis of the classification. This dichotomy has led to the introduction of a number of terms to describe various fractions and sub-fractions of carbohydrate.

Sugars

The term "sugars" is conventionally used to describe the mono and disaccharides. "Sugar", by contrast, is used to describe purified sucrose as are the terms "refined sugar" and "added sugar"

Extrinsic and intrinsic sugars

These terms had their origin in a United Kingdom (UK) Department of Health committee in 1989 (8), which was looking at the question of sugars in the diet. The terms were developed to help the consumer choose between what were considered to be healthy sugars and those which were not. Intrinsic sugars were defined as sugars occurring within the cell walls of plants, i.e. naturally occurring, while extrinsic sugars were those which were usually added to foods. Because lactose in milk is also an extrinsic sugar, an additional phrase "non-milk extrinsic sugars" was developed. These terms have not gained wide acceptance either in the UK or other countries in the world. There are no current plans to measure these sugars separately in the diet nor to incorporate their use into food tables.

Complex carbohydrates

This term was first used in the McGovern report, "Dietary Goals for the United States" in 1977. The term was coined largely to distinguish sugars from other carbohydrates and in the report denotes "fruit, vegetables and whole-grains". The term has since come to be used to describe either starch alone, or the combination of all polysaccharides. It was used to encourage consumption of what were considered to be healthy foods such as whole-grain cereals, etc., but becomes meaningless when used to describe fruit and vegetables which are low in starch.

Furthermore, it is now realised that starch, which is by any definition a complex carbohydrate, is variable metabolically with some forms being rapidly absorbed and having a high glycemic index and some being resistant to digestion. The term "complex carbohydrate" has encompassed, at various times, starch, dietary fibre and non-digestible oligosaccharides. As a substitute term for starch, however, it would seem to have little merit and, in principle, it is better to discuss carbohydrate components by using their common chemical names.

Available and unavailable carbohydrate

A major step forward conceptually in our understanding of carbohydrates was made by McCance and Lawrence in 1929 with the division of dietary carbohydrate into available and unavailable. In an attempt to prepare food tables for diabetic diets they realised that not all carbohydrates could be "utilised and metabolised", i.e. provide the body with "carbohydrates for metabolism". Available carbohydrate was defined as "starch and soluble sugars" and unavailable as "mainly hemicellulose and fibre (cellulose)". This concept proved useful, not the least because it drew attention to the fact that some carbohydrate is not digested and absorbed in the small intestine but rather reaches the large bowel where it is fermented.

It suggests that the site of digestion or fermentation in the gut of carbohydrate is of overriding importance. However, it is misleading to talk of carbohydrate as "unavailable" because some indigestible carbohydrate is able to provide the body with energy through fermentation. There are many properties of carbohydrate of which digestibility and fermentability are only two. A more appropriate substitute for the terms "available" and "unavailable" today would be to describe carbohydrates as either as glycemic (i.e. providing carbohydrate for metabolism) or non-glycemic, which is closer to the original concept of McCance and Lawrence.

Starch

Resistant starch

One of the major developments in our understanding of the importance of carbohydrates for health in the past twenty years has been the discovery of resistant starch. Resistant starch is defined as "starch and starch degradation products not absorbed in the small intestine of healthy humans". The main forms of resistant starch are physically enclosed starch, e.g. within intact cell structures (RS1), some raw starch granules (RS2) and retrograded amylose.

Modified starch

The proportions of amylose and amylopectin in a starchy food is variable and can be altered by plant breeding. Techniques using genetic engineering are rapidly emerging, enabling starches to be produced for specific purposes by genetically modifying the crop used for their production. High amylose corn starch and high amylopectin (waxy) corn starch have been available for a long time, and display quite different functional as well as nutritional properties. High amylose starches require higher temperatures for gelatinisation and are more prone to retrograde and to form amylose-lipid complexes. Such properties can be utilised in the formulation of foods with low glycemic index and/or high resistant starch content.

Physical modifications of starches include pregelatinisation and partial hydrolysis (dextrinisation). Chemical modification is mainly the introduction of side groups and cross-linking or oxidation. These modifications may be used to decrease viscosity and to improve gel stability, mouthfeel, appearance and texture, and resistance for heat treatment. The application of modified starches as fat replacers is another important area. Some modified starches may be partly resistant to digestion in the small intestine, thereby adding to resistant starch.

Fibre

Dietary fibre

The original description of dietary fibre by Trowell in 1972 was "that portion of food which is derived from cellular walls of plants which is digested very poorly by human beings". This is not an exact description of any carbohydrate in the diet but is more a physiological concept. It was linked by Burkitt and Trowell to the etiology of a number of "Western diseases" and on the basis of this a hypothesis relating fibre to health was developed. The use of the term has, however, caused many difficulties over the years because of controversies regarding definition. Moreover, the proposal that there are a number of dietary fibre deficiency disorders is an oversimplification and needs to be modified now in the light of new knowledge of diet and disease.

The main components of dietary fibre are derived from the cell walls of plant material in the diet and comprise cellulose, hemicellulose and pectin (the non-starch polysaccharides). Lignin, a non-carbohydrate component of the cell wall is also often included. Dietary fibre is a term which is felt to be valuable for the consumer who looks upon this as a healthy component of the diet. At the present time there is no consensus as to which components of carbohydrate should be included as dietary fibre and different authors have variously included non-starch polysaccharides and resistant starch.

More recently it has been suggested that non-digestible oligosaccharides should also be included. Dietary fibre has also been defined by method. While there is general agreement that the non-starch polysaccharides are the principal part of dietary fibre there is currently no consensus as to whether other components should be included in this term. It has been suggested that the use of the term dietary fibre be gradually phased out. Its widespread use and popularity with the consumer has made this difficult in practice and the term has been useful in nutrition education and product development.

Soluble and insoluble fibre

These terms developed out of the early chemistry of non-starch polysaccharides which showed that the fractional extraction of these polysaccharides could be controlled by changing the pH of solutions. They proved very useful in the initial understanding of the physiological properties of dietary fibre, allowing a simple division into those which principally had effects on glucose and lipid absorption from the small intestine (soluble) and those which were slowly and incompletely fermented and had more pronounced effects on bowel habit (insoluble).

However, the separation of soluble and insoluble fractions is not chemically very distinct being dependent on the conditions of extraction. Moreover, the physiological differences are not, in fact, so distinct with much insoluble fibre being rapidly and completely fermented while not all soluble fibre has effects on glucose and lipid absorption.

Analysis of Dietary Carbohydrate

Mono- and disaccharides

They can be analysed specifically by enzymatic, gas-liquid chromatography (GLC) or high performance liquid chromatography (HPLC) methods. Depending on the food matrix to be analysed, extraction of the low molecular weight carbohydrates in aqueous ethanol, usually 80% (v/v), may be advisable before analysis. The enzymatic procedures are based on specific, highly purified enzymes and have been instrumental in providing means of specific and precise analysis of individual carbohydrates in mixtures without a large investment in instrumentation.

Enzymatic methods are still preferable when one single carbohydrate is to be analysed, e.g. glucose, as the end point of starch analysis. When several different monosaccharides are to be determined simultaneously, HPLC or GLC methods are preferable. HPLC systems using sensitive amperometric detectors are gaining in popularity over GLC, in that the derivatisation necessary before the GLC determination is avoided.

Polyols

Polyols are usually determined by GLC using alditol acetate derivatives. HPLC methods are also available.

Oligosaccharides

Oligosaccharides can also be determined by GLC or HPLC methods. These methods work well for purified preparations, but in complex foods or diets, enzymatic hydrolysis and determination of liberated monosaccharides is an alternative for specific determination. Malto-oligosaccharides are recovered as "starch" if not extracted before starch analysis.

Separation of oligosaccharides

By definition, polysaccharides have 10 or more monomeric units, and Oligosaccharides less than 10. Analytically, separation is based on solubility in aqueous ethanol, usually around 80% (v/v). The alcohol solubility of carbohydrates, however, is dependent not only on the degree of polymerisation (DP), but also on the molecular structure. For instance, highly branched carbohydrates may be soluble in 80% ethanol in spite of a DP considerably higher than 10. In practice, therefore, the separation of Oligosaccharides from polysaccharides is empirical and does not provide an exact division based on DP.

Starch

Quantitative analysis of starch in foods by most current methods is based on enzymatic degradation and specific determination of liberated glucose. Nutritionally, starch can be divided into glucogenic ("available") and resistant starch, which is not absorbed in the small intestine. Resistant starch is poorly soluble in water and methods aiming at a total starch analysis employ an initial 2M potassium hydroxide (KOH) or dimethylsulfoxide solvent (DMSO) treatment to disperse crystalline starch fractions that would otherwise remain unhydrolysed.

Methods for measuring resistant starch are still in their infancy and have not yet been tested in formal collaborative studies. They aim at simulating normal starch digestion in the small intestine. A key step is to mimic the normal disintegration of the food which occurs during chewing. One method uses a standardised milling/homogenisation technique, whereas others employ standardised chewing by volunteers. Both approaches have been evaluated against human ileostomy experiments with a limited number of food matrices.

Non-starch polysaccharides (NSP)

The determination of NSP is based on the following steps: (a) degradation of starch by enzymatic hydrolysis after solublisation, (b) removal of low molecular weight carbohydrates, including starch hydrolysis products, (c) hydrolysis of the NSP to their constituent monomers, and (d) quantitative determination of those monomers. The acid hydrolysis step is a critical one, and it has to be designed as an optimal balance between complete hydrolysis and destruction of the liberated monomers.

The most widely-used method today for specific determination of the liberated monomers is GLC with alditol acetate derivatives. HPLC detection is an alternative gaining in popularity. Calorimetric determination is still preferred for uronic acids, which are derived mainly from pectic substances. A calorimetric method is also available for total NSP. Fractions of NSP, such as cellulose and non-cellulosic polysaccharides, can be separated by using sequential extraction and hydrolysis methods. For instance, cellulose is not hydrolysed by dilute (1-2M) sulphuric acid, unless it has first been dispersed in concentrated acid.

Dietary Fibre

Three methods for dietary fibre analysis have undergone extensive testing in recent years, including collaborative studies satisfactory enough for official approval of bodies such as the AOAC International (Association of Official Analytical Chemists) and the Bureau Communautaire de Reference (BCR) of the European Community:

— The enzymatic, gravimetric AOAC methods of Prosky and co-workers, and subsequently Lee and co-workers.

— The enzymatic-chemical methods of Englyst and co-workers.

— The enzymatic-chemical method of Theander and co-workers.

The enzymatic-gravimetric AOAC methods are derived from methods aiming at simulating the digestion in the human small intestine to isolate an undigested residue as a measure of dietary fibre. This residue is corrected for associated ash and protein. Since no DMSO or KOH dispersion is used, starch that resists the amylases used in the assay will remain as a fibre component. Since the sample has to be milled, and since a heat-stable amylase (termamyl) is used at a temperature close to 100°C, physically enclosed starch (RS1) and resistant starch granules (RS2) will not be included.

Retrograded amylose (RS3) that is included is the main form of resistant starch (RS) in processed foods. Lignin, a non-carbohydrate component of the dietary fibre complex is also included, as well as some tannins. These components are a very small proportion of most foods but can be substantial in some unconventional raw materials or special "fibre" preparations. The Englyst method measures the NSP specifically, either as individual monomeric components by GLC (or HPLC) or colourimetrically as reducing substances (total NSP).

Accordingly, DMSO is used initially to ensure a complete removal of starch, and lignin is not determined. The difference between estimates with the gravimetric methods and the Englyst method is mainly due to resistant starch and lignin. The Uppsala method employs hydrolysis conditions and GLC determination of monomers in a similar way as in the Englyst method. However, DMSO is not employed for starch dispersion, and a gravimetric estimate of lignin (Klason lignin) is added to obtain the dietary fibre. The Uppsala method and the gravimetric AOAC methods give very concordant results.

Intake of Carbohydrates

Trends in the supply and intake of carbohydrates can be studied by four principal approaches:

— Production

— Food balance sheets

— Household surveys

— Individual assessments

Food production statistics, which are available from FAO for every country in the world and for every crop, are useful for examining trends in consumption. From these data it can be seen that the major sources of carbohydrate in the human diet are:

— Cereals

— Root crops

— Sugar crops

— Pulses

— Vegetables

— Fruit

— Milk products

Trends over the last 20-30 years indicate growth in world production of cereals, sugar cane, vegetables and fruit. On the other hand, production of root crops, pulses and sugar beet has changed little on a world basis. Marked decreases have actually been seen in pulse production in some countries in Asia, and in root crop production in Europe. This suggests a change in food preference away from roots and pulses and towards cereals. Examination of eating habits in a number of countries indicates that this is the case.

Since root crops are an excellent source of carbohydrate, there is concern about this downward trend in production. Populations continue to grow in most parts of the world and, overall, food production would seem to be keeping pace with population growth. Increased production is due to improved agricultural practices rather than increased crop area, the major reason for increases being greater use of fertiliser. There are, however, specific countries where this is not happening. For the entire continent of Africa, cereal production is inadequate.

A major question is how much more improvement and efficiency in production can be achieved, and whether the amount of carbohydrate will be sufficient for the world's population in the future. Projections for future growth suggest problems ahead, particularly in Africa. Both food balance information and results from individual assessments are used to determine carbohydrate intakes. Food balance data is intended to describe food available for consumption. It is unlikely to do so because it does not include home production, which is variable from country to country, and may be considerable in some developing countries.

As a reflection of food consumed, food balance data is questionable, since it does not include food wasted or spoiled, or used for purposes other than human food, the proportion of which may change from year to year. As a result, food balance data for individual countries has failed to demonstrate the changes in consumption of carbohydrates which are seen using individual surveys.

Data from individual surveys also have limitations. Surveys are carried out by a variety of methodologies. While each has advantages and disadvantages, all suffer from a degree of under-reporting. This can be intentional or involuntary, most likely due to individuals forgetting food items or not describing foods thought to be undesirable. There is also the failure to record data or the altering of actual diets. The difference, then, between food balance data and individual assessments, for energy and nutrient intakes, is not only the form of wastage and spoilage on the food balance side of the equation, but also the under-reporting on the individual intake side.

True food intakes therefore lie somewhere between food balance and individual intake estimates. Another major problem is the varied carbohydrate terminology used in different countries. Many countries express total carbohydrate 'by difference', rather than as carbohydrate analysed directly, and this results in overestimates of the percent energy derived from carbohydrate. There is also a great variety in terms used to describe simple sugars, such as "sugars", "sugar", "refined sugar", "added sugar", "sucrose", and "sugars minus lactose".

Often there is no description of what is being reported. There is a need to standardise the terminology for carbohydrate and its components in individual surveys and a need for consistency in both reporting and the description of the terms used. In spite of terminology difficulties, it is possible to gain a picture of carbohydrate intakes and trends. As a percent of energy, total carbohydrate ranges from about 40% to over 80%, with the developed countries, such as those in North America, Western Europe and Australia at the low end of the range, and developing countries in Asia and Africa at the high end.

Starch accounts for 20%-50% or more of energy where the total carbohydrate intake is in the high range. Sugars account for 9%-27% of energy intake; where total carbohydrate is high, sugar intake is generally low. Where data are available, intake of carbohydrate as a percent of energy is higher for children than for adults. Trends in consumption indicate a falling carbohydrate intake in developed countries until the last two decades.

During that time some increase has been noted as fat intakes fall. The major sources of carbohydrate are cereals, representing over 50% of all carbohydrate consumed in both developed and developing countries, with sugar crops the next major source, followed by root crops, fruits, vegetables, pulses and milk products. In some of the developing countries much of the carbohydrate is derived from a single food source such as rice, cassava or maize. Carbohydrate foods are an important vehicle for protein, micronutrients and other food components, like phytochemicals, which have important benefits for health.

Individual food sources vary, however, in the provision of these components. A single food source of carbohydrate is therefore undesirable and populations whose diets are primarily based on a single food can suffer from micro-nutrient deficiencies due to lack of variety. It is important, therefore, that a number of different carbohydrate sources be consumed and efforts should be made to encourage a wide variety of carbohydrate foods.

Data on intake of sources of sugars is only available for developed countries. These data show similar proportions of sugars are derived from cereal products, milk products and beverages, among these countries. There is some variation in the proportions derived from fruit and confectionery, with the UK consuming less fruit and higher amounts of confectionery than countries such as the United States and Australia.

Intakes of non-starch polysaccharides range from about 19g/day in some countries in Europe and North America, to nearly 30g/day in rural Africa. Cereals are again the major source of this component. Data on intake of dietary fibre, determined by methods such as that of the AOAC (Association of Official Analytical Chemists) and the older Southgate method, are about 15-20 g/day for North America, Europe and Australia, to 25-40 g/day for countries in Asia and Africa.

Physiological Effects of Carbohydrates

Carbohydrates have a wide range of physiological effects which may be important to health, such as:

- Provision of energy
- Effects on satiety/gastric emptying
- Control of blood glucose and insulin metabolism
- Protein glycosylation
- Cholesterol and triglyceride metabolism
- Bile acid dehydroxylation
- Fermentation
    - Hydrogen/methane production
    - Short-chain fatty acids production
    - Control of colonic epithelial cell function
- Bowel habit/laxation/motor activity
- Effects on large bowel microflora

Energy Source

Dietary carbohydrates have by convention been given an energy value of 4 kcal/g (17 kJ/g), although where carbohydrates are expressed as monosaccharides, the value of 3.75 kcal/g (15.7 kJ/g) is used. It is now clear, however, that a number of carbohydrates are only partly or not at all digested in the small intestine and are fermented in the large bowel to short chain fatty acids. These include the non-digestible oligosaccharides, resistant starch and non-starch polysaccharides. The process of fermentation is metabolically less efficient than absorption in the small intestine and these carbohydrates provide the body with less energy.

In light of a new understanding of the digestion and metabolism of carbohydrate and developments in methodology, the energy value of all carbohydrates in the diet should be reassessed and more accurate energy factors assigned to each group or subgroup. There are a number of potential approaches to accomplish this. These include the classic calorimetry experiments similar to those first undertaken by Atwater, as well as human balance studies and ileostomy recovery experiments.

Knowledge of the chemistry of individual carbohydrates allows a prediction to be made regarding their digestion or fermentation, and an energy value to be assigned. In vitro models of fermentation can be constructed and from these the fermentation stoichiometry can be deduced. Studies using stable isotope tracer techniques may also be of value. While the energy yield of carbohydrate delivered to the colon will vary according to the extent of colonic fermentation (or the assumptions made in the model used), there may be an argument for assigning a single energy value to all such carbohydrate. Published studies suggest that a caloric value of about 2 kcal/g (8 kJ/g) would be a reasonable average figure for carbohydrate which reaches the colon. While individual carbohydrates will have different values, in the range of 1-2 kcal/g, these differences are unlikely to be of importance to health.

Satiety

The possibility of controlling hunger, satiety and food intake by altering the type of carbohydrate in food has intrigued a number of investigators. At present the variability of the findings and the lack of understanding of a clear relationship to physiologic parameters thought to be involved in the regulation of food intake limit practical application of this approach. It is unlikely that controlling a single dietary component, such as the type of sugar or starch, will lead to significant changes in the amount of food consumed. Also, compensation for small dietary changes made in one meal may often be seen at a subsequent meal. A better approach to controlling hunger and increasing satiety is likely to be associated with changes in the composition of the total diet.

Glucose and Insulin

The digestion of dietary carbohydrates starts in the mouth, where salivary a-amylase initiates starch degradation. The starch fragments thus formed include maltose, some glucose and dextrins containing the 1,6-a -glycosidic branching points of amylopectin. The a -amylase degradation of starch is completed by the pancreatic amylase active in the small intestine.

Dietary disaccharides, as well as degradation products of starch, need to be broken down to monosaccharides in order to be absorbed. This final hydrolysis is accomplished by hydrolases attached to the intestinal brush-border membrane, referred to as "disaccharidases". Disaccharidase deficiencies occur as rare genetic defects, causing malabsorption and intolerance of the corresponding disaccharide.

Glucose and galactose are transported actively against a concentration gradient into the intestinal mucosal cells by a sodium dependent transporter (SGLT 1). Fructose undergoes facilitated transport by another mechanism

(GLUT 5). Fructose taken together with other sugars (as in naturally fructose-containing foods) is better absorbed than fructose alone. When delivered to the circulation, the absorbed carbohydrates cause an elevation of the blood glucose concentration.

Fructose and galactose have to be converted to glucose mainly in the liver and therefore produce less pronounced blood glucose elevation. The extent and duration of the blood glucose rise after a meal is dependent upon the rate of absorption, which in turn depends upon factors such as gastric emptying as well as the rate of hydrolysis and diffusion of hydrolysis products in the small intestine.

Insulin is secreted as a response to blood glucose elevation but is modified by many neural and endocrine stimuli. Insulin secretion is also influenced by food related factors, especially by the amount and the amino acid composition of dietary proteins. Insulin has important regulatory functions in both carbohydrate and lipid metabolism and is necessary for glucose uptake by most body cells.

Lactose

Lactose, a β -linked disaccharide of glucose and galactose, is the principal sugar in milk. At birth, lactase activity is high in the brush-border of the small bowel of infants, but declines after weaning so that most populations of the world have low activity in adult life. The exceptions are Caucasian peoples and some other population groups in whom the majority retain a high lactase activity throughout life.

During the years since 1980, there has been a major change in the way lactose absorption is viewed and a resultant shift away from the concept that lactose "malabsorption" is a pathological state. Low mucosal lactase activity in adults is the norm throughout most of the world. However, such a state usually allows the drinking of modest quantities of milk spaced throughout the day without adverse symptoms. Milk consumption is therefore now being encouraged in many areas of the world because of its value as a source of protein, calcium and riboflavin.

Fermented milk products, which have lower lactose content and contain enzymes and microorganisms that can assist in lactose digestion, are better tolerated than milk. Technology exists to reduce the lactose level in foods and this should be taken into consideration when milk is included as food aid. Cheese, however, has almost no lactose. Lactose which is not digested, passes into the colon where it is fermented. In some individuals this causes lactose intolerance, the term used to describe the clinical symptoms of abdominal discomfort, flatulence and diarrhoea, associated with the ingestion of lactose containing foods by persons with low lactase activity.

It also occurs as a transient phenomenon when the intestinal mucosa is injured following acute infection in children and in protein-energy malnutrition. It is also found in adults, particularly in association with coeliac disease and tropical sprue. In these conditions, lactose malabsorption is said to be "secondary" to intestinal mucosal disease. A small proportion of the Caucasian population also exhibits low lactase activity and lactose intolerance.

Protein Glycosylation

The non-enzymatic glycation of proteins is dependent on the concentration of glucose and fructose in blood and the half-life of the protein. The initial reaction is between the monosaccharide and the amino group of an amino acid, usually lysine, to form a Shiff base which undergoes rearrangement and formation of Amadori products. As the reaction progresses, increasingly complex Maillard products are formed with the eventual production of Advanced Glycation End-products or AGEs which are associated with irreversible loss of protein function.

The extent of glycation of specific proteins, such as Haemoglobin Ale in diabetics serves as an indication of medium term control of blood glucose. Examples of functional changes induced by glycation include lens proteins in the eye with resultant cataract formation, increased microvascular complications, abnormal fibrin network formation and impaired fibrinolysis. These changes are most clearly seen in diabetic patients.

Lipids and Bile Acids

There has been concern that a substantial increase in carbohydrate-containing food at the expense of fat, might result in a decrease in high-density lipoprotein and a corresponding increase in very low-density lipoprotein and triglycerides in the blood. However, there is no evidence that this happens when the increase in carbohydrates occurs as a result of increased consumption of vegetables, fruits and appropriately processed cereals over prolonged periods.

Polysaccharides like oat β -glucan, guar gum and those from psyllium have been repeatedly shown to lower serum cholesterol levels in those with elevated levels, with little change if serum levels are normal. Proposed mechanisms include impaired bile acid and cholesterol reabsorption through physical entrapment in the small intestine, or inhibitory effects on cholesterol synthesis by products of lower bowel fermentation, particularly propionic acid. Not all fermentable polysaccharides are effective, however, and recent studies have indicated that neither oligosaccharides nor resistant starch have a significant effect on serum lipids in young normolipidemic subjects.

Fermentation

Fermentation is the colonic phase of the digestive process and describes the breakdown in the large intestine of carbohydrates not digested and absorbed in the upper gut. This process involves gut microflora and is unique to the colon of humans because it occurs without the availability of oxygen. It thus results in the formation of the gases hydrogen, methane and carbon dioxide, as well as short chain fatty acids (SCFA) (acetate, propionate and butyrate), and stimulates bacterial growth (biomass).

The gases are either absorbed and excreted in breath, or passed out via the rectum. The major products of such fermentation are the SCFA which are rapidly absorbed and metabolised by the body. Acetate passes primarily into the blood and is taken up by liver, muscle and other tissues. Propionate is a major glucose precursor in ruminant animals such as the cow and sheep, but this is not an important pathway in humans. Butyrate is metabolised primarily by colonocytes and has been shown to regulate cell growth, and to induce differentiation and apoptosis.

Bowel habit

It has long been known that non-starch polysaccharides are the principal dietary component affecting laxation. This occurs through increases in bowel content bulk and a speeding up of intestinal transit time. The extent of the effect depends on the chemical and physical nature of the polysaccharides and the extent to which they are fermented in the colon. Fermentable polysaccharides stimulate increases in microbial biomass in the colon, resulting in some increase in fecal weight, but not to the extent of non-fermentable polysaccharides. The latter are not significantly degraded in the colon and become consituents of the stool. In so doing, they hold water and produce a marked increase in fecal weight. Similarly, resistant starch can increase fecal weight, but this again depends on the extent of fermentation.

Microflora

Carbohydrate which is fermented stimulates the growth of bacteria in the large gut. This is a generalised effect which leads to an increase in the total number of bacteria or biomass. When bacterial growth occurs, the microflora synthesise protein actively from preformed amino acids and peptides as well as some denovo synthesis using ammonia as the source of nitrogen. The additional biomass is excreted in feces and is one of the mechanisms whereby carbohydrate influences bowel habit. The increased biomass excretion is accompanied by increased nitrogen excretion.

The efficiency of conversion of carbohydrate to biomass is determined principally by the type of substrate, the rate of breakdown and the transit time through the large intestine. One of the more significant developments in recent years with regard to the gut microflora has been the demonstration that specific dietary carbohydrates selectively stimulate the growth of individual groups or species of bacteria. An example of this is the effect of fructo-oligosaccharides on the growth of bifidobacteria. The importance of bifidobacteria is that they may be one of the major contributors to colonisation resistance in the colon, thereby protecting the host from invasion by pathogenic species. Foods which selectively stimulate the growth of gut bacteria are known as prebiotics.

Carbohydrates in Health Maintenance

Dietary Carbohydrates

While the amount of carbohydrate required to avoid ketosis is very small (about 50 g/day), carbohydrate provides the majority of energy in the diets of most people. There are many reasons why this is desirable. In addition to providing easily available energy for oxidative metabolism, carbohydrate-containing foods are vehicles for important micronutrients and phytochemicals. Dietary carbohydrate is important to maintain glycemic homeostasis and for gastrointestinal integrity and function.

Unlike fat and protein, high levels of dietary carbohydrate, provided it is obtained from a variety of sources, is not associated with adverse health effects. Finally, diets high in carbohydrate as compared to those high in fat, reduce the likelihood of developing obesity and its co-morbid conditions. An optimum diet should consist of at least 55% of total energy coming from carbohydrate obtained from a variety of food sources.

The consultation agreed that when carbohydrate consumption levels are at or above 75% of total energy there could be significant adverse effects on nutritional status by the exclusion of adequate quantities of protein, fat and other essential nutrients. In arriving at its recommendation of a minimum of 55% of total energy from carbohydrate, the consultation realised that a significant percentage of total energy needs to be provided by protein and fat, but that their contribution to total energy intakes will vary from one country to another on the basis of food consumption patterns and food availability.

Maintenance of Energy Balance

In adults, it is important that the amount of energy ingested be matched to the amount of energy expended. Maintenance of energy balance is important in order to avoid obesity and its associated co-morbidities such as diabetes and cardiovascular disease. Positive energy balance and obesity occur when total energy intake exceeds total energy expenditure, regardless of composition of the excess energy. However, the composition of the diet can affect whether and to what extent positive energy balance occurs.

The composition of the diet can also affect the ability to maintain energy balance. In particular, diets containing at least 55% of energy from a variety of carbohydrate sources, as compared to high fat diets, reduce the likelihood that body fat accumulation will occur. Substantial data suggest that diets high in fat content tend to promote consumption of more total energy than diets high in carbohydrates. This effect may be due to the low energy density of high carbohydrate diets, since total volume of food consumed appears to provide an important satiety cue.

There are no data to suggest that different types of carbohydrates differentially affect total energy intake. In addition to affecting the chance of having excess energy available, the composition of the diet also affects the proportion of excess energy that will be stored as body fat. The body has a large fat storage capacity and excess dietary fat is stored very efficiently in adipose tissue. Alternatively, the body's capacity to store carbohydrate is limited and excess carbohydrate is not efficiently stored as body fat. Instead, excess carbohydrate tends to be oxidised, leading to indirect fat accumulation via reductions in fat oxidation.

Excess fat and carbohydrate were previously thought to be equally flattening. This was due to the assumption that de novo lipogenesis was a commonly used pathway for disposal of excess carbohydrate. The available data suggest, however, that this process occurs rarely in human subjects and only in situations of appreciable carbohydrate overfeeding. In most usual circumstances, accumulation of body fat via de novo lipogenesis is quantitatively very low. While noting the low overall contribution of de novo lipogenesis to body fat accumulation, it should be noted that de novo lipogenesis is increased with insulin resistance and with extremely high consumption of sucrose or fructose.

Physical Activity

Maintenance of energy balance is dependent both on energy intake and energy expenditure. Maintaining regular physical activity greatly reduces the likelihood of creating positive energy balance, regardless of the composition of the diet. There is agreement that the combination of a high carbohydrate diet and regular physical activity is the optimal arrangement to avoid positive energy balance and obesity. The increased energy needs of physical activity can be supplied by carbohydrate or fat. The importance of carbohydrate in the diet becomes more critical as the amount and intensity of physical activity increases.

In many developing countries, the major challenge is to meet daily energy needs created by high levels of daily physical labour. In such cases, any combination of carbohydrate and fat which provides sufficient energy is to be encouraged. Many countries recommend increasing leisure time physical activity. While increased physical activity would clearly increase energy needs, these do not create needs for specific macronutrients. Rather, the optimum diet identified above is considered sufficient to provide for such physical activity.

There is substantial evidence that supplemental carbohydrate can improve performance for the elite endurance-trained athlete. A high carbohydrate diet during a few days preceding an endurance event, carbohydrate loading, a high carbohydrate pre-event meal and carbohydrate supplementation in the form of carbohydrate-containing beverages have all been shown to enhance performance during long-distance cycling and running. There is, however, no evidence that such carbohydrate supplementation would improve performance for the majority of people who engage in recreational physical activity of lower intensity and duration. On the other hand, carbohydrate intake following exercise can help to quickly replenish depleted glycogen stores.

Impacts on Behaviour

It has been suggested that food intake could have important effects on behaviour. While providing breakfast to children who do not typically eat breakfast can increase cognitive performance, it is less clear that the overall composition of the diet can affect behaviour. It has been suggested that sugar consumption leads to hyperactivity in children. However, an extensive review of the literature in this area concluded that there is no evidence to support the claim that refined sugar intake has any significant influence on either behaviour or" cognitive performance in children.

Because glucose is an essential fuel for the central nervous system, carbohydrate has also been suggested to play a role in memory and cognitive function. While there appears to be a relationship between glucose levels and memory processing, the clinical significance of this relationship remains unclear.

Energy and nutrient needs are increased in pregnancy and lactation, and the primary challenge for pregnant women is to meet these increased energy needs in order to ensure healthy offspring. It has been observed that where variety in the food supply is low and carbohydrate intake is high, a low birth weight is more common. This raises concerns about the adequacy of high carbohydrate diets to meet the energy and nutrient needs of pregnancy when food variety is limited. Energy and nutrient needs should be met by consumption of a wide variety of carbohydrate foods. There is also some concern about excessive fat intake in pregnancy since it may be associated with risk of obesity in the mother.

In many countries, infants receive 45-55% of energy from fat through breastmilk or formulas and 35-45% of energy from carbohydrate. While specific reductions in fat intake are not recommended below the age of two years, infants in many countries consume lower fat diets. This does not present a problem as long as energy requirements are fulfilled. From the age of two and on, the optimum diet (at least 55% of total energy from a variety of carbohydrate sources) should be gradually introduced. During the first four to six months of life, exclusive breast feeding is recommended as this tailors the concentration of lactose to the maturing neonatal and infant gut, particularly while colonic microflora and pancreatic amylase production are developing.

For infants fed on formula, the carbohydrate and other nutrient components should usually mimic breast milk to the extent possible and in accordance with standards of the Codex Alimentarius. Carbohydrate digestion in the neonate and young infant is significantly influenced by both gastrointestinal maturation and the chemical nature of the carbohydrate ingested. The establishment of colonic microflora is responsible for colonic carbohydrate scavenging, converting any carbohydrate entering the colon into short chain fatty acids. Any disturbances or inappropriate development of this microflora (incorrect infant formula, antibiotics, infection) leads to colonic carbohydrate overloading and diarrhoea.

Lactose from dairy products can be a major source of carbohydrate for young children. In addition, milk represents an excellent source of high quality protein, calcium, and riboflavin. In most populations, even those with low lactase activity, milk can be ingested in small amounts, especially after meals with dilution by co-ingestion. Fermented dairy products can be valuable items in the diet of most people irrespective of intestinal lactase status. Often the transition from childhood to adulthood is associated with changes in dietary pattern.

In developing countries, children frequently consume very high carbohydrate intakes from a single or a small number of sources, while adults have greater variety. In such cases, the adult diet is preferred. In developed countries, on the other hand, surveys indicate that children have higher intakes of carbohydrate from more sources than adults. In those countries, the diet consumed by children would seem to be more beneficial. In both situations, at least 55% of carbohydrate energy from a variety of sources is the optimum.

Individualisation of carbohydrate intake is necessary for elderly populations. Elderly individuals in many countries are at risk as regards both malnutrition and obesity. Food intake patterns can be altered by changes in taste perception, chronic disease and medication use. While a high carbohydrate diet is recommended for prevention of weight gain and obesity, it should be recognised that some individuals may need diets higher in energy density (e.g. fats) in order to prevent malnutrition. Optimising intake of carbohydrate to minimise glucose intolerance in later life is a consideration in countries where such intolerance is a problem.

Carbohydrate and Disease

Carbohydrates may directly influence human diseases by affecting physiological and metabolic processes, thereby reducing risk factors for the disease or the disease process itself. Carbohydrates may also have indirect effects on diseases, for example, by displacing other nutrients or facilitating increased intakes of a wide range of other substances frequently found in carbohydrate-containing foods. Evidence of associations between carbohydrates and diseases comes from epidemiological and clinical studies. There are relatively few examples in which direct causal links between carbohydrates and diseases have been proven.

Obesity

The frequency of obesity has increased dramatically in many developed and developing countries. This is of profound public health importance because of the clearly defined negative effect of obesity, especially when centrally distributed, in relation to diabetes, coronary heart disease and other chronic diseases of lifestyle. Genetic and environmental factors play a role in determining the propensity for obesity in populations and individuals. Lack of physical activity is believed to contribute to the increasing rates of obesity observed in many countries and may be a factor in whether an individual who is at risk will become overweight or obese.

High carbohydrate foods promote satiety in the short term. As fat is stored more efficiently than excess carbohydrate, use of high carbohydrate foods is likely to reduce the risk of obesity in the long term. Much controversy surrounds the extent to which sugars and starch promote obesity. There is no direct evidence to implicate either of these groups of carbohydrates in the etiology of obesity, based on data derived from studies in affluent societies.

Nevertheless, it is important to reiterate that excess energy in any form will promote body fat accumulation and that excess consumption of low fat foods, while not as obesity-producing as excess consumption of high fat products, will lead to obesity if energy expenditure is not increased. While high carbohydrate diets may help reduce the risk of obesity by preventing overconsumption of energy, there is no evidence to suggest that the macronutrient composition of a low energy diet influences the rate and extent of weight loss in the treatment of obese patients.

Non-insulin Dependent Diabetes Mellitus (NIDDM)

High rates of NIDDM in all population groups are associated with rapid cultural changes in populations previously consuming traditional diets, and also with increasing obesity, especially when centrally distributed. Although the precise mode of inheritance has not been established, there is no doubt that genetic factors are involved. Certain populations appear to have a strong predisposition to the development of NIDDM to the extent that in some groups about half the adult population have the disease.

Within all populations a family history of NIDDM is an important predisposing factor. Diet and lifestyle-related conditions which may lead to obesity will clearly influence the risk of developing NIDDM in populations and individuals who are susceptible to this condition. Foods rich in non-starch polysaccharides and carbohydrate-containing foods with a low glycemic index appear to protect against diabetes, the effect being independent of body mass index. In terms of disease prevention, it is not possible on the basis of current data to distinguish the relative merits of different types of non-starch polysaccharides.

Some epidemiological evidence suggests particular benefit of appropriately processed cereal foods, while other epidemiological and clinical studies suggest benefits of non-starch polysaccharide from legumes and pectin-rich foods. Thus, avoiding obesity and increasing intakes of a wide range of foods rich in non-starch polysaccharide and carbohydrate-containing foods with a low glycemic index offers the best means of reducing the rapidly increasing rates of NIDDM in many countries.

Consuming a wide range of carbohydrate foods is now regarded as acceptable in the nutritional management of people who have already developed NIDDM. It has been suggested that between 60 and 70 per cent of total energy should be derived from a mix of mono-unsaturated fatty acids and carbohydrates. Carbohydrates should principally be derived from a wide range of appropriately processed cereals, vegetables and fruit, with particular emphasis on those foods which have a low glycemic index. The goal to achieve and maintain ideal body weight remains paramount, ensuring that foods high in fat which might predispose to obesity are not encouraged, even though they might have a low glycemic index.

Sucrose and other sugars have not been directly implicated in the etiology of diabetes and recommendations concerning intake relate primarily to the avoidance of all energy-dense foods in order to reduce obesity. Most recommendations for the management of diabetes permit modest (30-50 g/day) intakes of sucrose and other added sugars in the diabetic dietary prescription provided these are:

a) consumed within the context of total energy allowance;

b) nutrient-dense foods and foods rich in non-starch polysaccharides are not displaced; and,

c) they are incorporated as part of a mixed meal.

In some populations where fat intake is relatively low and sucrose intake high, a reduced intake of sucrose may be considered in the diabetic dietary prescription. Increased meal frequency under iso-energetic conditions does not, in the long term, appear to be associated with any alteration in glycemic control. This suggests that personal preference is the key determinant of meal frequency, provided that body weight and daily (as well as long-term) glycemic control are not adversely influenced.

Special diabetic food products are not generally recommended and fructose is not regarded as having any particular merits as a sweetener when compared with other added sugars. However, low-energy beverages containing alternative non-nutritive sweeteners may be useful for people with diabetes. Dietary factors have not been conclusively shown to be risk factors for insulin-dependent diabetes and the key advice concerning carbohydrates in the management of this condition concerns distribution of intake of carbohydrates during the day. Carbohydrate intake needs to be regularly distributed and balanced with injected insulin. The general principles of the diabetic dietary approach to non-insulin dependent diabetes may also be applied to those with insulin-dependent diabetes.

Cardiovascular Disease

Many genetic and lifestyle factors are involved in the etiology of coronary heart disease and influence both the atherosclerotic and thrombotic processes underlying the clinical manifestations of this disease. Dietary factors may influence these processes directly or via a range of cardiovascular disease risk factors. Obesity, particularly when centrally distributed, is associated with an appreciable increase in the risk of coronary heart disease. There is also evidence implicating specific nutrients and, in particular, high intakes of some saturated fatty acids appear to be important promoters of coronary heart disease.

On the other hand, there is increasing evidence of a strong protective effect by a range of antioxidant nutrients. Increasing carbohydrate intake can assist in the reduction of saturated fat and many fruits and vegetables rich in carbohydrates are also rich in several antioxidants. Cereal foods rich in non-starch polysaccharides have been shown to be protective against coronary heart disease in a series of prospective studies. There is no evidence for a causal role of sucrose in the etiology of coronary heart disease.

The cornerstone of dietary advice aimed at reducing coronary heart disease risk is to increase the intake of carbohydrate-rich foods, especially cereals, vegetables and fruits rich in non-starch polysaccharide, at the expense of fat. Among those who are overweight or obese it is more important to reduce total fat intake and to encourage the consumption of the most appropriate carbohydrate-containing foods. There has been concern that a substantial increase in carbohydrate-containing food at the expense of fat, might result in a decrease in high-density lipoprotein and an increase in very low-density lipoprotein and triglycerides in the blood.

There is, however, no evidence that this occurs when the increase in carbohydrates results from increased consumption of vegetables, fruits and appropriately processed cereals, over prolonged periods. Certain non-starch polysaccharides (for example β-glucans) have been shown to have an appreciable effect in lowering serum cholesterol when consumed in naturally occurring foods, or foods which have been enriched by purified forms, or even when fed as dietary supplements. Such polysaccharides may be used in the management of patients with existing hypercholesterolemia but their role, if any, in the prevention of coronary heart disease remains to be established.

Less information is available concerning the role of carbohydrates in other cardiovascular diseases. Plant foods are good sources of potassium and reducing the sodium to potassium ratio may help to reduce the risk of hypertension. Limited data suggest a protective effect of vegetables and fruit in cerebrovascular disease. There has been considerable debate in many developed countries which have high rates of coronary heart disease regarding the age at which children should start to reduce fat intake towards the recommended level for adults.

Clearly children require an adequate intake of energy for growth, and it is important that this does not include an excessive intake of carbohydrates at a very young age. It is generally accepted that dietary carbohydrate should gradually be increased and fat reduced after the age of two years, so that by the age of five years children should have reached a diet in the range of that recommended for adults. This advice should, of course, include the key dietary guidelines for children and adolescents, which suggest that nutritional adequacy should be achieved by eating a wide variety of foods and that energy intake should be adequate to promote growth and development, and to reach and maintain desirable body weight.

Cancer

Diet is widely regarded as important in the etiology of colorectal cancer with meat and fat considered the primary risk factors, and fruit, vegetable and cereal foods considered to be protective. Cancer is a disease associated with well-recognised genetic abnormalities and for colorectal cancer in particular, defects in a number of genes have been clearly defined. These genes mostly code for proteins responsible for the control of either cell growth, cell-to-cell communication or DNA repair. They are mainly oncogenes or tumor suppressor genes.

For the development of colorectal cancer an individual must acquire several of these genetic abnormalities in the same cell. The acquisition of gene defects in somatic cells is thought to be through DNA damage and a resultant failure of the DNA repair system (or of apoptosis). Dietary carbohydrate is thought to be protective through mechanisms involving arrest of cell growth, differentiation and selection of damaged cells for cell death (apoptosis). This is probably achieved primarily through the action of butyric acid which is formed in the colon from fermentation of carbohydrates such as resistant starch and non-starch polysaccharides. Such carbohydrates are found mostly in cereals, fruit and vegetables.

The process of fermentation may protect the colorectal area against the genetic damage that leads to colorectal cancer through other mechanisms which include:

a) the dilution of potential carcinogens;

b) the reduction of products of protein fermentation through stimulation of bacterial growth;

c) pH effects;

d) maintenance of the gut mucosal barrier; and,

e) effects on bile acid degradation.

These mechanisms, however, are much less well-established. Carbohydrate staple foods are a source of phytoestrogens which may be protective for breast cancer. Cancer risk is increased for the obese. This applies especially to cancers of the breast and uterus. However, this is a general effect of total energy intake and not specifically of carbohydrates. Dietary carbohydrates do not have a known role in the etiology of lung, breast, stomach, prostate, pancreas, oesophagus, liver or cervical cancers. There is, however, some evidence that there is an increased risk of ovarian cancer in women with mild galactosemia.

Gastrointestinal Diseases

Intakes of non-starch polysaccharides and resistant starch are the most important contributors to stool weight. Therefore, increasing consumption of foods rich in these carbohydrates is a very effective means of preventing and treating constipation, as well as haemorrhoids and anal fissures. Bran and other cereal sources containing non-starch polysaccharide also appear to protect against diverticular disease and have an important role in the treatment of this condition. Obesity is an important risk factor for gallstones. High intakes of carbohydrate may facilitate the colonisation of bifidobacteria and lactobacilli in the gut and thus reduce the risk of acute infective gastrointestinal illnesses.

Dental caries

The incidence of dental caries is influenced by a number of factors. Foods containing sugars or starch may be easily broken down by a-amylase and bacteria in the mouth and can produce acid which increases the risk of caries. Starches with a high glycemic index produce more pronounced changes in plaque pH than low glycemic index starch, especially when combined with sugars. However, the impact of these carbohydrates on caries is dependent on the type of food, frequency of consumption, degree of oral hygiene performed, availability of fluoride, salivary function, and genetic factors. Prevention programmes to control and eliminate dental caries should focus on fluoridation and adequate oral hygiene, and not on sucrose intake alone.

Other Conditions

There are a number of inherited conditions having significant implications for restricted dietary carbohydrate intake in infants and children. These include rare conditions such as galactosemia, fructose intolerance, a wide range of glycogen storage diseases, sucrose deficiencies and monosaccharide transport deficiencies. Though rare in incidence, their early detection and careful dietary management is important if severe handicap or pathology is to be avoided.

Role of the Glycemic Index in Food Choices

Carbohydrate foods often contain vitamins and minerals plus other compounds, such as phytochemicals and antioxidants, which may have health implications. Consuming a wide variety of carbohydrate foods is therefore recommended as this is more likely to be a nutritionally adequate diet with the health benefits commonly ascribed to carbohydrate foods. Food choice depends not only on nutrition and health considerations but also on factors such as local availability, cultural acceptability and individual likes and needs. There is no one measure which can be used to guide food choices in all cases.

The chemical composition of foods (e.g. fat, sugars, dietary fibre content) should be an important factor influencing food choice. However, simply knowing the chemical nature of the carbohydrates in foods, for example, does not reliably indicate their actual physiologic effects. Foods which are good choices in some situations may not be the best choices in others. Likewise, foods which are poor choices in some situations may be good choices in others. Two indices of carbohydrate foods based on their physiologic functions have been proposed.

A recently suggested satiety index measures the satiety value of equal energy portions of foods relative to a standard, which is white bread. The factors which control food intake are complex and satiety needs to be distinguished from satiation. Nevertheless, investigation of satiety indices of foods is considered an interesting area of future research, which, if validated, may aid in the selection of appropriate carbohydrate foods to promote energy balance. A more established index is the glycemic index which can be used to classify foods based on their blood glucose raising potential.

The glycemic index is defined as the incremental area under the blood glucose response curve of a 50g carbohydrate portion of a test food expressed as a percent of the response to the same amount of carbohydrate from a standard food taken by the same subject. A number of different methods have been used to calculate the area under the curve. For most glycemic index data, the area under the curve has been calculated as the incremental area under the blood glucose response curve (IAUC), ignoring the area beneath the fasting concentration. This can be calculated geometrically by applying the trapezoid rule. When a blood glucose value falls below the baseline, only the area above the fasting level is included. Sample data are shown in Table 1.

Table 1: Sample blood glucose responses to the ingestion of 50g carbohydrate

Minutes	0	15	30	45	60	90	120	IAUC
Standard #1	4.3	6.3	7.9	5.3	4.1	4.6	4.9	114
Standard #2	4.0	6.0	6.7	5.5	5.3	5.0	4.2	155
Standard #3	4.1	5.8	8.0	6.5	5.9	4.8	3.9	179
Test Food	4.0	5.0	5.8	5.4	4.8	4.2	4.4	93

The portion of food tested should contain 50g of glycemic (available) carbohydrate. In practice, glycemic carbohydrate is often measured as total carbohydrate minus dietary fibre, as determined by the AOAC method. Since this method does not include RS 1 and RS2 when they are present, they will be mistakenly included as glycemic carbohydrate.

Blood glucose response is normally measured in capillary whole blood. Plasma glucose can be used to determine the glycemic index and gives similar values. However, capillary blood is preferred because it is easier to obtain, the rise in blood glucose is greater than in venous plasma and the results for capillary blood glucose are less variable than those for venous plasma glucose. Thus, differences between foods are larger and easier to detect statistically using capillary blood glucose.

Either white bread or glucose can be used as the standard food. The GI values obtained if white bread is used are about 1.4 times those obtained if glucose is the standard food. Other standard foods could be used, but to enable comparison with data in the literature, the GI of the new standard food relative to standardised white bread or glucose should be established.

Blood glucose responses vary considerably from day-to-day within subjects. Thus, to obtain a representative mean response to the standard food, it is recommended that the standard food be repeated at least three times in each subject. This is illustrated by the data in table 2, which is typical for normal subjects. The standard food was repeated three times giving IAUCs of: 114, 155 and 179. The mean ± SD IAUC is 149 ± 33 and the coefficient of variation (100 × SD/mean) is 22%. For this subject, the GI of the test food = 100 × 93/149 = 62. To determine the GI of the food, the tests illustrated in table 2 would be repeated in six more subjects and the resulting GI values averaged. Normally, the GI for more than one food would be determined in one series of tests, for example, each subject might test four foods once each and the standard food three times for a total of seven tests in random order on separate days. Subjects are studied on separate days in the morning after a 10-12 ho overnight fast. A standard drink of water, tea or coffee should be given with each test meal.

Glycemic Properties of Foods

Starchy foods with a low GI are digested and absorbed more slowly than foods with a high GI. Some factors that influence glycemic properties of foods are listed in table 2

Table 2: Food factors influencing glycemic responses

Amount of carbohydrate	
Nature of the monosaccharide components	
	Glucose
	Fructose
	Galactose
Nature of the starch	
	Amylose
	Amylopectin
	Starch-nutrient interaction
	Resistant starch
Cooking/food processing	
	Degree of starch gelatinization
	Particle size
	Food form
	Cellular structure
Other food components	
	Fat and protein
	Dietary fibre
	Antinutrients
	Organic acids

Calculation of Glycemic Index

The GI can be applied in a detailed fashion to mixed meals or whole diets by calculating the weighted GI value of the meal or diet. For example, the way to calculate the GI of a meal containing bread, cereal, sucrose, milk and orange juice is shown in Table 3. Using this type of calculation, there is a good correlation between meal GI and the observed glycemic responses of meals of equal nutrient composition. Blood glucose responses are also influenced by the amount of carbohydrate in the meal. To compare the expected glycemic load of meals with different carbohydrate contents, a non-linear adjustment can be applied, but this has only been tested in normal subjects.

Table 3: Calculation of the glycemic index of meals

Food	Grams Glycemic Carbohydrate	Proportion of total Glycemic Carbohydrate	Food Glycemic Index	Meal Glycemic Index
Bread	25	0.342	100	34.2
Cereal	25	0.342	72	24.6
Milk	6	0.082	39	3.2
Sucrose	5	0.068	87	5.9
Orange juice	12	0.164	74	12.1
Total	73	—	—	80.0

For detailed application of the GI, a value of the GI for every food in the diet or meal needs to have been assigned (for many foods the value has to be estimated). The accuracy of the calculation depends upon the accuracy of the GI values ascribed to foods, which may vary from place to place due to local factors such as variety, cooking, processing, etc. Foods particularly prone to such variation include rice, potatoes and bananas.

Use of the Glycemic Index

The glycemic index can be used, in conjunction with information about food composition, to guide food choices. For practical application, the glycemic index is useful to rank foods by developing exchange lists of categories of low glycemic index foods, such as legumes, pearled barley, lightly refined grains (e.g. whole grain pumpernickel bread, or breads made from coarse flour), pasta, etc. Specific local foods should be included in such lists where information is available (e.g. green bananas in the Caribbean and specific rice varieties in Southeast Asia).

In choosing carbohydrate foods, both glycemic index and food composition must be considered. Some low GI foods may not always be a good choice because they are high in fat. Conversely, some high GI foods may be a good choice because of convenience or because they have low energy and high nutrient content. It is not necessary or desirable to exclude or avoid all high GI foods.

Physiologic and Therapeutic Effects

Meals containing low GI foods reduce both postprandial blood glucose and insulin responses. Animal studies suggest that incorporating slowly digested starch into the diet delays the onset of insulin resistance. Some epidemiologic studies suggest that a low GI diet is associated with reduced risk of developing non-insulin diabetes in men and women. Clinical trials in normal, diabetic and hyperlipidemic subjects show that low GI diets reduce mean blood glucose concentrations, reduce insulin secretion and reduce serum triglycerides in individuals with hypertriglyceridemia.

In addition, the digestibility of the carbohydrate in low GI foods is generally less than that of high GI foods. Thus, low GI foods increase the amount of carbohydrate entering the colon and increase colonic fermentation and short chain fatty acid production. This has implications for systemic nitrogen and lipid metabolism, and for local events within the colon.

Guidelines for Carbohydrate Food Choices

Principles

The principles are:

— To acknowledge the sociocultural context, lifestyle and stage of life-cycle, in food carbohydrate choice;

— To give preference to food choices rather than to nutrient goals in carbohydrate food choices, and in so doing:

  a) Use food categories as a guide to chemically defined carbohydrate type.

  b) Use numbers of portions (serving sizes) of foods from designated food categories in order to provide semi-quantitative food-based advice. This may imply that meal frequency would need to increase in some cultures, because the accommodation of enough carbohydrate food in the course of the day, without an excessive amount on any one occasion, requires more frequent servings and consumption;

— To appreciate that many of the world's health problems are associated with inadequate carbohydrate intake, and potentially also associated with inappropriate carbohydrate intake;

— To ensure the acceptability and practicality of any recommended change in carbohydrate food intake;

— To acknowledge that there may be unintended consequences involved in carbohydrate food intake change, and also to ensure that risks involved in dietary changes from traditional diets is considered;

— To monitor the intake of carbohydrate foods and, wherever possible, of chemically defined carbohydrate components of those foods in relation to health issues.

— To ascertain whether food carbohydrate choices encourage biodiversity and are sustainable.

Nutrient Goals

The minimum amount of carbohydrate in the human diet that is needed to avoid ketosis is of the order of 50 g/day in adults. Beyond this, additional energy needs are best met by nutrient-dense carbohydrate foods. There must, of course, be adequate intakes of protein (with essential amino acids) and essential fatty acids from fat. Moderate intake of sugar-rich foods can also provide for a palatable and nutritious diet.

Food Goals

There are a number of approaches to translating nutrient recommendations to food goals:

— Recommending the total weight of food groups to be consumed. Various national food guides have suggested quantities of specific foods to be consumed, such as fruits and vegetables, and pulses, nuts and seeds.

— Examining sources of carbohydrate foods in various diets, particularly diets which have desirable total carbohydrate intakes or from countries where the incidence of lifestyle diseases is low. Recommendations can then be made on the basis of intakes.

— Examining major food groups which contain carbohydrate foods and recommending numbers of servings of those food groups. Numerous countries around the world, both developed and developing, have produced food guides with such groupings, and considering the level of agreement that exists for carbohydrate as a percent energy, these food guides are remarkably consistent in their advice.

— Examining indices which exist to describe various physiological properties of carbohydrate-containing foods, such as glycemic index values and values from other indices which are presently being developed.

On the basis of the above approaches, and taking into account the principles for carbohydrate food choice, the following recommendations can be made:

— A variety of foods should provide the carbohydrate in the diet, not a single or small number of sources.

— Cereals, roots, pulses, fruit and vegetables are all components of a healthy diet throughout the world.

— Cereal foods or root crops, where this is the main staple, should provide the major source of carbohydrate energy.

— Intake of fruits and vegetables (including potatoes in developed countries) should be high. As well as being a valuable source of carbohydrate, fruit and vegetables are an important source of antioxidant vitamins and other food components.

— Consumption of pulses, nuts and seeds should be encouraged. While this group often represents only a small amount of carbohydrate energy, these foods are a good source of protein and micronutrients. They should be consumed with cereals to optimise protein quality.

— At least small quantities of milk products are desirable, even when low lactase activity exists, since these are a good source of protein and micronutrients.

These recommendations apply to all individuals over the age of two years, with adjustments as necessary for growth and the increased demand of pregnancy and lactation.

Achieving goals for intake of carbohydrates does not ensure nutritional adequacy. Carbohydrate foods provide a range of nutrients and other substances essential for health in addition to energy. It is therefore essential to consume a variety of foods in order to derive the full benefits of a high carbohydrate diet. Nutrients from foods require monitoring. For example, in Iran carbohydrate foods include vegetables (250 g/day), fruits (210 g/day), pulses (20 g/day) and cereals (wheat at about 250 g/day and rice 110 g/day, uncooked).

Wheat has mostly been consumed as bread with traditional pastries being festive foods. Recently, consumption of the latter has increased substantially with potential reduction of nutritionally useful food groups traditionally accompanying bread. This may not significantly impact on total carbohydrate intake, but could influence nutritional adequacy. Traditional methods of food preparation and preservation facilitate food choice variety, and promote nutritional benefits. Alteration of traditional practices may compromise such benefits. Ongoing monitoring may be required to guard against nutritional inadequacy.

Planners and policy makers need to:

1. Recognize dietary goals of at least 55% of total energy from a variety of carbohydrate sources.
2. Recognize the extent of change necessary in order to meet these goals
3. Understand that there must be a gradual transition in meeting new dietary guidelines and that new terminology will need to be gradually accepted.
4. Consider the effects of economic and cultural factors in achieving dietary goals.
5. Develop clear guides about the types and quantities of food recommended.
6. Develop methods to monitor food consumption to meet dietary goals.

Primary producers and processors need to:

1. Consider how existing and new technologies can be used to help meet dietary goals regarding the quantity and nutritional properties of food carbohydrates, as well as levels of micronutrients and other desirable food components.
2. Provide foods, such as breakfast cereals and snack foods that are high in NSP, low in energy density, and with a low glycemic index.
3. Increase the availability and convenience of fruits and vegetables.
4. Provide appropriate information to the consumer on food labels.

References

Asp, N.-G. (1995). Classification and methodology of food carbohydrates as related to nutritional effects. *American Journal of Clinical Nutrition* 61(4(S)):930S -937S.

Asp, N-G. (1996). Dietary carbohydrates: classification by chemistry and physiology. *Food Chemistry* 57:9-14.

Department of Health. (1989). *Dietary sugars and human health.* Her Majesty's Stationery Office, London.

Englyst, H.N. and Hudson, G.J. (1996). The classification and measurement of dietary carbohydrates. *Food Chemistry,* 57(1): 15-21.

FAO. (1980). Carbohydrates in human nutrition, a Joint FAO/WHO Report. *FAO Food and Nutrition Paper 15,* Rome.

Southgate, D.A.T. (1991). *Determination of food carbohydrates.* Elsevier Science Publishers, Ltd., Barking.

5

Energy and Protein Requirements

The requirements for energy and protein of an individual are defined in the following terms:

Energy. The energy requirement of an individual is the level of energy intake from food that will balance energy expenditure when the individual has a body size and composition, and level of physical activity, consistent with long-term good health; and that will allow for the maintenance of economically necessary and socially desirable physical activity. In children and pregnant or lactating women the energy requirement includes the energy needs associated with the deposition of tissues or the secretion of milk at rates consistent with good health.

Protein. The protein requirement of an individual is defined as the lowest level of dietary protein intake that will balance the losses of nitrogen from the body in persons maintaining energy balance at modest levels of physical activity. In children and pregnant or lactating women, the protein requirement is taken to include the needs associated with the deposition of tissues or the secretion of milk at rates consistent with good health.

All requirement estimates refer to needs persisting over moderate periods of time. The corresponding intakes may be referred to as "habitual" or "usual", to distinguish them from intakes on a particular day. However, as a matter of convention and convenience they are expressed as daily rates (of intake). However, there is no implication that these amounts must be consumed each day.

There are important physiological differences between the requirements for energy and protein, as defined above. For energy, it is usually considered that once the level of body weight and physical activity has been fixed and the appropriate growth rate defined, there is only one level of intake at which energy balance can be achieved; in consequence, this becomes that individual's requirement for energy. Even if some degree of adaptation is possible, it is likely that this range is fairly narrow. If the intake is either above or below the requirement, defined in this way, a change in body energy stores is to be expected unless energy expenditure is correspondingly altered. If such changes in expenditure do not occur, the energy store, mainly in the form of adipose tissue, will increase when the intake exceeds requirement and decrease when it is below requirement. It is clearly contrary to experience to suppose that for each individual there is one fixed setpoint for body weight and adipose tissue mass compatible with health. In fact, we recognize that for any individual there is probably a range of acceptable body weights, and this also applies to the individuals in a group. However, if the imbalance is too great, or continues over long periods, the resulting changes in body weight and composition can be detrimental to function and health. Consequently, there may be risks associated with intakes either above or below actual requirements.

For protein, the requirements of individuals are also expressed in terms of the amount of dietary protein needed to prevent losses of body protein and to allow, where appropriate, for desirable rates of deposition of protein during growth and pregnancy. In this respect, the requirements of the individual for both protein and energy are analogous. However, in contrast to energy, if more protein is ingested than is needed for metabolic purposes, essentially all the excess is metabolized and the end-products are excreted, since protein is not stored in the body in the way that energy is stored in adipose tissue. Furthermore, again in contrast to energy, no detrimental effect has been identified with intakes of protein moderately above the actual requirement. For an individual, the range between the intake that is just sufficient to compensate for losses (or permit growth) and intakes that are associated with harmful effects is therefore wide. The individual's requirement is thus defined as the lower end of this range. This fundamental biological difference between energy and protein has important consequences for describing the distribution of requirements among the individuals of a group.

Individuals and Groups

After defining the requirements of an individual, the next step is to extend this definition to those of a group. Estimates of requirements are derived from measurements on individuals. Actual measurements on people of the same sex and of similar age, body size, and physical activity are in practice grouped together to give the average energy or protein requirement of that set of people, together with a measure of their variability. These results are then used to predict the requirements of other individuals or collections of individuals who have the same characteristics, but on whom measurements have not been made. Such a collection of similar individuals may be referred to as a *class*.

The characteristics of the class are that obvious factors that may affect requirements—age, sex, weight, etc.—have been matched. However, in spite of the matching there remain many unknown factors producing variation between individuals, so that there is a distribution of requirements within the class. Changes in the variables that characterize the class will involve a change in the average requirement and therefore a change in the position of the distribution.

For a class of similar individuals, the descriptor of energy requirements is the *average* of the individual requirements, without specific provision for the known individual variation in requirement.

Common Considerations to the Estimation of Energy and Protein Requirements

Adaptation

In the context of nutrition, a working definition of adaptation might be: "a process by which a new or different steady state is reached in response to a change or difference in the intake of food and nutrients". The word "new" includes the individual who is responding to a change, e.g., when a subject in a balance study moves from a high- to a low-protein intake. The word "different" is appropriate when comparisons are made between individuals or groups who are habitually exposed to different environmental or nutritional conditions. It might be useful to distinguish these as short-term and long-term adaptations.

Three general points are important in relation to both types of adaptation.

(*a*) The concept of a steady state is relative. No one is ever in an absolutely steady state, either in body weight or in nitrogen balance. During the day, when food is being eaten, nitrogen balance is positive; during the night, in the absence of food, it becomes negative. Over 24 hours these fluctuations even out. The situation is precisely analogous to the oxygen debt incurred during strenuous exercise, which is made up during rest. The time-scale over which a state may be considered steady will vary for different functions. It is also clear that in biological systems no so-called steady state is completely stable. This is illustrated by the slow changes in body composition and function that occur as adults age.

(*b*) Adaptations may be of fundamentally different kindsmetabolic, biological/genetic, and social/behavioural. The response to a change in protein intake is one of the best worked out examples of a metabolic adaptation, but even in this case we do not know the limits of human adaptability.

Reduction in physical activity as a consequence of a reduced energy intake could be regarded as a behavioural adaptation, with both good and bad effects. For example, the energy intakes of children may be adequate to support satisfactory growth rates, but only at the expense of a reduction in total energy expenditure, notably through less physical activity. The result is to impair the child's capacity for exploration and play, and hence his mental, functional, and social development.

It has often been suggested that a decrease in body size might be an advantageous adaptation to shortage in food supply. The determinants of body size are complex and depend upon both environmental and genetic factors. In developing countries large numbers of children are small for their age compared with the NCHS standards). This reduction in linear growth ("stunting") is the result of environmental factors because it is reversible under favourable conditions. Whether it represents a handicap is hotly debated. It has also been suggested that selection might occur in favour of those who are genetically smaller and hence have lower needs. A full analysis of the significance of differences in body size would have to take account also of the secular changes that have occurred in a number of countries and of the established relationship between the heights of parents and their children.

(*c*) It follows from these considerations that adaptation implies a range of steady states and it is impossible to define a single point within the range that represents the "normal". Different adapted states will carry different advantages and penalties. A decision about which is optimal or preferable can only be made in the light of a particular set of values. If the criteria are life expectancy and freedom from disease in the early years of life, then perhaps the nutritional state in industrialized societies might be preferred to that of developing countries, but there may be other criteria of optimal functional capacity.

The same point was made in the report of the 1971 Committee, in an apt quotation from Atwater & Benedict : "One essential question is, what level is most advantageous? The answer to this must be sought not simply in metabolism experiments and dietary studies, but also in broader observations regarding bodily and mental efficiency and general health, strength and welfare".

The concept of a range of adapted states, each with advantages and disadvantages, produces a dilemma: it implies respect for different biological and cultural situations, but it may also encourage the acceptance of double standards and the endorsement of the *status quo*. To quote again from the report of the 1971 Committee: "when supplies are insufficient and purchasing power is low, consumption is likely to be less than requirements. In such circumstances 'what is' will not be 'what should be'.

Body Size

Body size is the major determinant of the absolute requirements for energy and protein. Variations in size are probably more significant quantitatively than the metabolic adaptations discussed in later sections. It is therefore necessary at the outset to define acceptable ranges of body size.

Children

Although the energy and protein requirements for the process of growth are relatively small compared with those for maintenance, except in the young infant, satisfactory growth is nevertheless a sensitive criterion of whether needs are being met. Therefore, the definition of satisfactory growth is the first step in estimating the requirements of infants and children. An example of the dilemma mentioned above is whether reference standards for the growth of children in industrialized countries should be accepted as universally relevant or whether local standards should be used. Children in many developing countries are smaller at birth than those in industrialized countries and grow at a slower rate during infancy and early childhood. The evidence suggests that in young children these differences are due primarily to environmental factors, including inadequate nutrition, and that genetic and ethnic factors are of lesser importance, so that young children of different ethnic groups should be considered as having the same or similar growth potential.

Even in a healthy privileged population there is a wide range of variation in the size of children. In such a population there is no indication that differences in size*per se* are related to health, wellbeing, or physiological function. However, in communities where children's growth is limited by environmental factors, there is evidence of an association between functional impairment and deficit in linear growth. In such situations, it is extremely difficult to separate the effects of undernutrition from those of other aspects of social deprivation. Therefore, it remains a matter for further research how far small size in children represents a handicap or an adaptation, whether minor limitations of genetic growth potential are harmful, and whether maximum growth is necessarily an indicator of optimal nutrition.

Nevertheless, the Consultation feels it desirable that the growth potential of children should be fully expressed, and estimates of energy and protein requirements should allow for this. Estimates of the requirements of children up to 10 years are therefore based on the reference growth standards published for international use by WHO, which are derived from the United States National Center for Health Statistics (NCHS). The use of this particular reference population is recommended on the basis of a number of criteria.

Surprisingly, in contrast to adults, there appear to be no recommendations for children concerning the ranges within which weight or height at any given age may be regarded as satisfactory. This is partly because information about the risk attached to given degrees of deficit is only just becoming available, and partly because children start with different birth-weights, and therefore their attained weights will continue for some time to be above or below the median. In epidemiological studies of childhood undernutrition it is conventional to accept —2 SD from the median as the cut-off point between "normal" and "malnourished", corresponding approximately to the 3rd centile or to 80% of the median for weight and

90% for height. Similarly, +2 SD in weight for height may be taken as a cut-off point for obesity.

In relation to growth, two further points must be taken into account. Previous committees have based their estimates of the daily requirements for growth on increments in body weight of the reference population over intervals of 3 months for infants below 1 year and intervals of 1 year for older children. This procedure assumes that growth occurs at the same rate from day to day. It must be recognized that this is not so, and that even in normal children free from infection growth occurs in spurts. As a result, the variability in weight gain over short periods such as one month is extremely high. For example, in two longitudinal studies from birth to 3–6 months, the coefficient of variation of weight gain over 4-week periods was approximately 37%. The reasons for this variability in the rate of weight gain are not clear. The data of Fomon (personal communication,1980) show that it is greater than the variability in energy intake. One factor, therefore, may be day-to-day fluctuations in physical activity. From the point of view of protein requirements, the phenomenon could be regarded as analogous, on a longer time-scale, to the metabolic differences that have been observed between the day, when food is consumed, and the night, when it is not. Studies on adults have shown that during the night there is a negative nitrogen balance that is cancelled out by a positive balance during the day. In children it has been suggested that growth occurs in spurts in relation to food intake.

The variability in growth would not affect requirements averaged over a period of time if it were a consequence simply of fluctuations in food intake, so that intake and rate of growth each day were exactly matched. It seems unlikely, however, that this is the full explanation. Since this exact matching does not happen, the effect of the variability in growth may be to increase the growth component of the requirement, as a reduction in growth over one period will have to be compensated by an increased growth rate later on. It is extremely difficult to estimate the quantitative effect of this variability in growth rate.

A second question that should be raised, although it cannot at present be answered, results from the fact that all allowances for growth are based on increments of body weight. It is conceivable that growth might be limited by special requirements of particular tissues; for example, the relative amounts of energy and protein needed to achieve a given gain in body weight might not be the same as the amounts needed to secure an appropriate increase in height. This is a subject on which further research is needed.

Adolescents

The desirable heights and weights of children over 10 years of age present special problems, because there is considerable variation between individuals and groups in the timing of the adolescent growth spurt, which starts at different calendar ages in boys and girls.

Furthermore, if children have been growing slowly from infancy, as happens in many developing countries, by 10 years of age the gap between their actual weight and their expected weight, based on that of adolescents in industrialized countries, will be very large. It is not known whether extra food at this stage can increase the extent and duration of the pubertal growth spurt.

For these reasons it is considered more realistic, after the age of 10 years, to relate requirements to the appropriate weight for height rather than weight for age. In order to maintain uniformity with the reference values for the earlier years of childhood, the standards published by WHO for height up to 18 years have again been chosen, but they do not include values of weight for height beyond 10 years. To provide these values the Consultation used data from a large sample of children measured in the United States of America earlier in this century.

Adults

Adult groups in various parts of the world show substantial differences in height, and height variations are also common within countries and races.

In general, there is no reason to suppose that adults of either short or tall stature have a health risk attributable to their stature, except perhaps in relation to pregnancy and childbirth, and therefore no attempt is made in this chapter to define the height of a healthy reference population of adults. However, body weight, when expressed in relation to height, does influence health, and a range of desirable or acceptable weights for height has been proposed.

Body Composition

Estimates of requirements based on body weight are an approximation, since they do not take account of differences in body composition, which will determine true requirements. In recent decades the emergence of methods for estimating some body components in living subjects has resulted in observations on several thousand people ranging from newborn infants to the very old. At birth the neonate averages 14% body fat, which rises to about 23% at 12 months and declines to 18% at 6 years of age. During this period girls have slightly more body fat than boys, and this difference becomes more pronounced after 6 years. During adolescence the difference in the body fat content between the sexes becomes strikingly accentuated and persists throughout adult life as shown by differences in the thickness of the subcutaneous fat layers. There is evidence that body composition is also influenced by genetic factors, since obesity tends to be familial and monozygous twins are more concordant in fatness than dizygous twins.

Adolescence is also characterized by a major sex difference in the rate of acquisition of lean weight. Boys show a rapid and sustained spurt in lean weight, whereas there is a modest acquisition of body fat in the early phase of puberty, followed by a decline. In contrast, girls have a smaller spurt in lean weight, but they acquire more body fat. In adolescent boys the time of the spurt in lean weight has been found to coincide with the most rapid growth in height, and to continue until 20–25 years of age, whereas in girls the pubertal increase in lean weight ceases by about 18 years, in keeping with the marked decrease in the rate of gain in stature after menarche. During the second decade of life boys thus double their lean weight, while the increase in girls is only 1.5-fold. The end result at maturity is a fat-free weight of about 60 kg in males averaging 70 kg total body weight, and 42 kg in females averaging 63 kg. The variability in lean weight is less than that of total body weight.

The adult years are characterized by a decline in lean body mass in both sexes, which becomes obvious by the age of 40; by 85 years the lean body mass has reached a value about three-quarters of that characteristic of the young adult. Simultaneous measurements of body nitrogen by neutron activation and body potassium as ^{40}K show that, with advancing age, more potassium than nitrogen is lost, implying that the potassium-rich muscle mass is especially reduced. The relative loss of potassium is about 10% between the sixth and eighth decades. Autopsy data confirm that subjects over 70 years of age have 40% less muscle than young adults, with a smaller reduction in the mass of visceral organs. Preferential loss of muscle with aging is also demonstrated by the decline in creatinine output and the fall in 3-methyl histidine output. This age-related loss of lean body mass is commonly accompanied by an increase in body fat. Consequently, in relation to body weight, the lean tissue content of the body declines with age, and this accounts in part for a progressive fall in basal metabolic rate in relation to body size. In the elderly there is a decrease in the proportion of skeletal mass, as well as muscle.

Within the lean body mass there are also differences in the proportion of various tissues at different ages. From birth to maturity the brain increases its mass 5-fold, the liver, heart, and kidneys, which are even more metabolically active, increase 10- to 12-fold, while muscle multiplies its mass by about 40-fold.

Physical Fitness and Functional Capacity

Estimates of requirements for energy and protein are based primarily on metabolic and balance studies of limited duration. However, the estimated requirements should be enough to maintain health and sustain optimal bodily function, including physical and mental fitness.

During growth, in addition to the increases in weight and height, there are marked functional changes. In boys, aerobic capacity and heart volume in relation to lean body mass increase up to the age of 14–15 years. The natural peak in functional capacity coincides with high levels of spontaneous physical activity, and estimates of requirements must allow for this. Conversely, in the fourth to the fifth decade of life aerobic capacity starts to decline, with decreasing physical activity and energy requirements.

Intakes of energy or protein above as well as below those needed for optimal function may be detrimental if they exceed the adaptive capacity of the organism. Excessive energy intakes lead to obesity, with reductions in cardiorespiratory efficiency, physical performance, and endurance.

It has already been suggested that stunting in linear growth may represent an adaptation that does not necessarily present any health hazard beyond early life. For example, cardiorespiratory function, physical performance, and muscular strength were found to be significantly better in stunted Tunisian children from a poor socioeconomic group than in children from affluent families, whose growth was closer to that of the standard in developed countries. Similarly, Italian children from poor families performed better in physical fitness tests than their counterparts from more prosperous families, in spite of their smaller size and lower habitual energy intakes. These findings suggest that habitual physical activity is a more important determinant of fitness than is body size *per se*. Nor do the effects of small stature necessarily carry penalties in adult life except for tasks requiring a particular body build and strength. Thus high aerobic capacity related to body weight was found in Indian miners with very low weights and heights compared with their counterparts in other countries. On the other hand, in another study in India low weight and height were significant handicaps for obtaining employment in agriculture. Therefore, while stunting in height is a rather sensitive marker of socioeconomically disadvantaged populations, the consequences need to be carefully evaluated for evidence of any functional handicap.

Expression of Requirements in Relation to Body Weight and Age

Relation to body weight

Protein. Within a given age range, the requirement for protein per kg of body weight is considered to be constant. Therefore, the *primary expression* of protein requirement is in grams of protein per kilogram. This principle applies to all ages, although absolute additions, in units of grams of protein per day, are made for pregnancy and lactation.

Energy. For energy the position is more complicated, because at a given age the main component of the energy requirement, the basal metabolic rate (BMR) varies not only with absolute body weight but also per kg. The total energy requirement per kg, therefore, cannot be taken as constant. Hence the *primary expression* of energy requirement is the total requirement per person, derived from that person's body weight.

With adults less than 60 years old the effect of age is relatively unimportant, since at a given weight the BMR decreases by only about 1% per decade. With children the change of BMR per kg with age is much greater—about 5% per year between 3 and 10 years. At present, we do not know to what extent this reflects an age-related change *per se*, or age-related changes in body weight. However, for this chapter the question is not of practical importance, since the BMR has not been used for estimating energy requirements in children below the age of 10 years.

The energy requirements of adolescents are treated in the same way as those of adults.

If at each age weight is the main determinant of requirements, the question then arises, what weight should be used? Within the acceptable range, many people will have weights that differ by 10% or more from the median. If the actual weight is used, it will tend to maintain the *status quo*. If the median of the reference range is used, the result will have a normative effect. The existing needs of those who are at the lower end of the weight range for age or height will be overestimated and the needs of those at the upper end will be underestimated. If the requirements so calculated are fulfilled, there will be a tendency for weight to move towards the median. This is what is meant by the term "normative", as used in this chapter. It will be for the user to choose the most appropriate body weight for calculating requirements, depending on the circumstances and his aims.

Interpretation of Tables of Requirements

The term "class" is introduced to take account of the variability that exists between apparently similar individuals. In practice this concept of a homogeneous class is not entirely realistic. For the purposes of tabulation it is necessary to group together individuals, according to age, weight for height, etc. The wider the ranges chosen, the less homogeneous such groups will be. The problem then is to decide on operationally useful ranges. No hard-and-fast guideline can be laid down. A large number of narrow ranges is more precise than a small number of wide ones, but also more complex. The ranges in many of the tables have therefore been chosen as a compromise between convenience and precision.

The variability of requirements within a group defined by any chosen range will clearly be greater than in a completely homogeneous class. This has implications for calculating the requirements of the group as a whole which are of great importance for the application of requirement estimates..

The requirements specified for each range in the tables represent those of the individual or homogeneous class whose weight for age or weight for height is at the mid-point of the range. Such estimates do not necessarily correspond with the average requirement of the group as a whole, since this will depend on the distribution of individuals within the group.

A particular problem arises if a group of people has a mean weight which is markedly different from the mid-point of the reference range. For example, the data collected by Eveleth & Tanner show that young Indonesian adults had on average a body mass index (BMI) of 19. It would be possible to specify the requirements of this group either on the basis of their actual average weight, or on the basis of the average acceptable weight for height (BMI 22), which would be about 15% greater. As in the case of the individual, the choice will depend upon whether or not the user considers the actual situation to be satisfactory.

There is, moreover, an additional complication. In such a group with a mean BMI of 19, there will be some individuals whose weight for height is below the limits of the acceptable range. concept of an acceptable range implies that some action is necessary to improve the situation of those who are outside it and are therefore in an unacceptable position. With a mixed group, some of whom are "inside" and some "outside", there are again two choices of action: to concentrate on individual outsiders, or to take measures which will affect the whole group.

Expression of age ranges

Conflicting methods are often used for expressing the ages or age ranges to which values of body weight, BMR, etc., refer. In some tables, e.g., those of the NCHS, the value for body weight opposite the figure for 5 years, for example, means the weight at 5.0 years. Since few children will be measured exactly on their birthday, such values are usually obtained by interpolation. In tables in other reports, the entry for weight at 5 years may in some cases signify the average weight of children between 4.5 and 5.5 years (mean age 5.0 years), in others the average weight of children from 5 to 6 years (mean age 5.5 years).

For reasons of clarity and to comply with ordinary usage, age range (e.g., 5–6 years) are specified in this chapter, because this is how children are usually classified, for example in school.

When a value for body weight or height is given, it represents the value at the mid-point of the range, obtained by interpolation from the NCHS standards. The range 5–6 starts at 5 years, up to but not including 6 years. To write 5–5.99 implies an unrealistic degree of precision. Another useful solution is to designate the ranges as 5+, 6+, etc..

When values (e.g., for BMR or protein requirement) are changing with age, it is clearly artificial to make dividing lines at particular ages at which abrupt changes are supposed to occur. In reality the values are continuous variables. If a more precise estimate is needed, it can be obtained by interpolation.

These are problems that relate more to applications than to the substance of this chapter, but it is necessary to outline them here in order to illustrate the different ways in which the figures and tables may be used.

Principles for the Estimation of Energy Requirements

General Considerations

The energy requirement was defined as the amount needed to maintain health, growth, and an "appropriate" level of physical activity. Using this definition it is impossible entirely to avoid value judgements on what is meant by health and appropriate activity. Values, and consequently decisions, may change under different conditions.

Energy needs are determined by energy expenditure. Therefore, in principle, as was recognized in the report of the 1971 Committee, estimates of requirements should be based on measurements of energy expenditure. This kind of information is difficult to obtain, and sometimes the only feasible approach is to estimate requirements from measurements of intake. If people are, on average, in a steady state, with appropriate body composition and levels of activity, measurements of their mean habitual intake will provide an estimate of their mean expenditure. The intention of the word "habitual" is to even out short-term fluctuations in intake, but it is not possible to define it with precision.

Requirements derived here are intended to apply to people who are healthy, and in general the effects of disease should be considered separately. However, it is recognized that in many populations this condition is unrealistic.

It cannot be assumed that observed levels of expenditure or intake always represent what is desirable for the maintenance of health. In developing countries actual intakes may be too low to allow for what was described in the report of the 1971 Committee as "leisure time" activity. Again, in an affluent society some people may be less physically active than is thought desirable to ensure cardiovascular health.

Components of Energy Requirement

In the great majority of cases the largest component of energy expenditure is the basal metabolic rate (BMR), which can be measured with accuracy under standardized conditions. In this chapter, therefore, the principle of calculating all components of total energy expenditure as multiples of the BMR has been adopted.

It is recognized that this principle, used for the sake of simplicity, is likely to involve some inconsistencies. The relationship of the energy cost of a given level of physical activity to BMR will be affected by the nature of that activity, whether static or dynamic; by the body weight, because of the different values of BMR per kg at different body weights; and by age, because of age-related changes in body composition and BMR.

Basal metabolic rate (BMR)

In any individual the BMR is determined principally by body size, body composition, and age. The relationships are complex; the BMR per unit weight varies with age, being higher in children and lower in the elderly. The BMR per unit weight also varies with weight: within a given age range, BMR per kg is higher in short and light individuals and lower in tall and heavy ones.

For practical purposes the most useful index of BMR is the body weight. In the report of the 1971 Committee, a table of weights and associated BMRs was given, taken from a paper by Talbot, based on measurements on 2200 children. For the present report, a more extensive set of measurements was compiled from the literature; these are representative of BMR in developed countries and of some data from developing countries.

The data base covers some 11 000 technically acceptable measurements on individuals of both sexes and all ages, who were considered to be healthy. It includes adults of different stature and of different weight for height, as well as individuals who fell within the designated range of "acceptable" weight for height. The data also include some children and adults who may have been on limited energy intakes before the BMR measurement. This may partly explain the lower BMRs in some groups.

Investigators may find the BMR of groups in their country differs from that predicted by the present general equations. This is to be expected, since in the data used for developing the current equations Indian subjects were found to have BMRs approximately 10% below the average and northern Europeans and North Americans tended to have higher values at equivalent age, height, and weight for height. These observations are not in themselves evidence for ethnic differences in BMR, so for the present the general equations have been maintained for all groups.

Growth

The energy cost of growth includes two components: the energy value of the tissue or product formed and the energy cost of synthesizing it. The total cost will therefore depend upon the composition of the product. The energy value is the heat of combustion, without the deductions for losses in urine and faeces which are allowed for by the Atwater factors. The average values for protein, fat, and carbohydrate are 5.7, 9.3, and 4.3 $kcal_{th}$ (24, 39, and 18 kJ) per g. Numerous estimates of the costs of synthesizing protein and fat have been derived from work on animals. The cost for protein is greater than for fat, even when fat is being synthesized from carbohydrate precursors. Except in the case of young infants and during lactation, the estimates of energy cost are not very critical, since human growth is a slow process, taking up a small proportion of the energy requirement. Moreover, since the composition of the tissue formed cannot be known accurately, and even the composition of breast milk is somewhat variable, it is only possible to make approximate estimates of the energy cost of growth.

In young children a rounded-off value of 5 $kcal_{th}$ (21 kJ) per g for the energy cost of growth has been widely accepted.In adults a higher figure is obtained for the energy cost of weight gain under different conditions. This may be because relatively more fat is being laid down.

The energy stored during pregnancy includes the energy laid down in the fetus, placenta, and uterus as well as the additional protein and fat stored in the mother. The composition of the tissue laid down varies at different stages of pregnancy, but since the overall extra cost is only some 10% of the total energy requirement, detailed computations, appropriate to the different stages of pregnancy, are not justified.

The energy requirement of the lactating woman must include the energy lost in milk. The daily amount should be taken as that produced by healthy well-nourished mothers. An additional allowance has to be made for the energy cost of producing the milk. As in the report of the 1971 Committee, the efficiency of conversion of food energy to milk energy has been taken as 80%.

In many societies women do not consume the extra amounts that seem to be needed to meet the energy requirements of pregnancy and lactation, and it has been suggested that there may be metabolic adaptations leading to increased efficiency. At present, however, there is no well documented evidence for this and further research is needed.

Physical activity

The level of physical activity must obviously be considered in detail when assessing energy needs. Some activities are essential for the individual and the community, and can be considered as economic activities which are life-sustaining. These are designated as *occupational* energy needs. Previous reports have given general estimates of the energy costs of light, moderate, and heavy activity. The difficulty of applying such figures in practice is recognized, since in many cases activity varies from day to day and from season to season.

The 1971 Committee included an allowance for "leisure-time" activities. The present Consultation attaches much importance to such activities, which are perhaps more appropriately termed *discretional*, as they are considered desirable for the wellbeing of the community and the health of the individual and the population.

Because of the wide range of variation in both occupational and discretionary activities, it is only possible to give examples typical of particular groups. The notes that follow provide some guidelines for the application of this approach.

1. *Occupational activities*. The traditional classification of work according to occupation is maintained in this chapter, but care must be taken to ensure that there is an adequate desrciption of the occupation. For example, farmers in affluent societies may be relatively sedentary compared with farmers in developing countries who are involved in very strenuous manual labour. The energy cost of travelling or walking to work should be considered as part of the essential energy needs. It must also be recognized that older children and women in developing countries commonly play a significant role in agriculture, in caring for livestock, and in looking after younger children. This is an important contribution to the economy and viability of the household, and energy should be allowed for these essential tasks.

2. *Discretionary activities*. There may be many benefits to societies from additional activities outside working hours. The requirement to cover them should not be considered as dispensable, since it usually contributes to the physical and intellectual wellbeing of the individual, household, or group. Such activities can be divided into three categories:

a. *Optional household tasks*. A number of optional tasks, such as working in the garden or repairing and improving the home, are an important part of family life. In estimating requirements, an energy allowance should therefore be made for adults for all these activities.

b. *Socially desirable activities*. A variety of socially constructive tasks, for example attending community meetings, games or festivals, or walking to health clinics or places of worship, require additional energy expenditure. In some developing countries, where people's main occupation involves a large expenditure of energy, there may be limits on the ability of members of the community to respond to a demand for activities of this kind. For children, additional energy is important as part of the normal process of development, for activities such as exploration of the surroundings, learning, and behavioural adjustments to other children and adults.

c. *Activity for physical fitness and the promotion of health*. At all ages physical fitness and wellbeing may depend on leisure-time exercise, and an allowance should be made for this even if it is recognized that at present many people in affluent societies do not expend enough energy in this way. Rather than reducing the estimate of energy requirement, it should be maintained to allow such people to become more physically active. It is, however, impossible to state precisely the desirable level and duration of extra activity.

In sedentary workers, cardiovascular responses to exercise are often inappropriate and muscular strength is limited. A small allowance for short periods of physical exercise at a relatively high rate would therefore be beneficial. Although not all authorities are agreed, there is in our view good evidence that for middle-aged men short regular periods of physical exercise at a relatively high rate may have a beneficial effect on cardiovascular risk.

Metabolic response of food

The increased oxygen uptake (so-called "specific dynamic action") after a meal depends on the nutrient composition of the food consumed, and the amount of energy ingested. The greater the energy demands of a subject, because of his size or physical activity for example, the greater the absolute rate of energy expenditure in digesting, absorbing, and storing the larger amounts of ingested nutrients. However, the measurement of the energy cost of these processes in the individual is not easy. It is difficult to separate the energy expended in excess of the basal rate after eating a meal, from the energy cost of the physical activity involved in sitting, eating, and digesting.

For the purpose of estimating energy expenditure, the practical solution is to measure the metabolic rate of individuals in the post-prandial state without limiting minor physical movement. The rate obtained in this way represents the resting metabolic rate. It is greater than the BMR because it includes the energy cost of metabolizing and digesting a meal, as well as the cost of increased muscle tone and minor physical activity. By combining the results of measurements made in the morning, afternoon, and evening, an average figure can be obtained which is an estimate of the resting metabolic rate.

Changes in Energy Requirements with Age

The most important component of energy expenditure, the basal metabolic rate, depends on the mass of metabolically active tissue in the body, the proportion of each tissue in the body, and the contribution of each tissue to the energy metabolism of the whole body. The changes in body composition with age, markedly affect energy requirements, since some organs of the body are much more metabolically active than others. In the neonate the brain comprises about 10% of the total body weight and may account for 44% of the total energy needs of the child under basal conditions. On the other hand, the energy needs for muscle metabolism at this time are very low because of the relatively small muscle mass. The liver is much more metabolically active than muscle, so that when the mass of muscle is reduced in the aging adult, the whole-body metabolic rate relative to lean tissue mass will also alter.

There are also altered activity patterns with age; children become progressively more active once they are able to crawl or walk and the physical activity patterns of adults are usually dominated by the nature of their work. If adults retire from work, this change in habits must be recognized in estimating their energy requirements, but it should not be assumed that the marked decline in activity that often occurs in the elderly is either inevitable or desirable. If energy intake declines with increasing inactivity, then an individual is much more likely to have a diet deficient in one of the essential nutrients. In determining the desirable minimum activities the Consultation suggests that the energy allowance considered appropriate for discretionary activities should be maintained throughout adult life and increased for those who have retired from work.

Sex Differences in Energy Requirements

The basal energy expenditure on a weight basis differs little between pre-adolescent boys and girls, but since there are differences in body weight and composition from the first few months of life, and different physical demands are made on boys and girls, their energy requirements are considered separately.

After maturity, men have a relatively greater muscle mass than women, which would tend to reduce their BMR when expressed in terms of lean body mass, since muscle has a low metabolic rate. However, the greater body fat content of women means that the observed BMR per unit total body weight is somewhat lower in women.

The energy demand for physical activity will often depend on the different types of employment for men and women. However, the heavy burden of agricultural work for women in rural communities must be taken into account. In addition, this chapter suggests that for both sexes there should be a minimum desirable energy requirement to allow the same amount of discretionary activity.

Variability in Energy Expenditure

In any assessment of the average requirement, both intra- and inter-individual variability must be recognized. The former results from short-term fluctuations in energy intake and expenditure. In the United Kingdom it has been found that measurements over a period of 2–3 weeks were needed in order to assess correctly the intakes of individuals. There are also short-term fluctuations in energy output, and it has been found that even under well controlled conditions, observations may have to be continued for several weeks before input and expenditure are balanced.

It is also generally recognized that in a group of apparently comparable people there is much inter-individual variation in habitual energy expenditure, and hence in requirement. In a number of selected studies, the measurement of total energy expenditure over a week indicates that the inter-individual variability of expenditure, in a specified group, has a coefficient of variation (CV) of about ± 12.5% on a body-weight basis.There is almost no information about the intra-individual variation in energy requirements in developing countries. Unpublished data from Papua New Guinea indicate a coefficient of variation of 10–16%.

Measurement of Energy Expenditure

Basal metabolic rate (BMR)

Direct measurement of the BMR demands close attention to detail and imposes artificial conditions on the subject —who should be in the post-absorptive state and at complete rest in a thermoneutral environment. In practice the BMR measured in this way is approximately equal to the energy expenditure of subjects during sleep. It is therefore considered valid to measure the BMR of individuals and to assign this energy cost to the time during which the subject is asleep. Anxiety is often cited as an important cause of increased energy expenditure, but direct estimates do not confirm this.

Physical activity

The different types of activity undertaken by an individual can be identified and the time spent in each activity measured. The energy cost of each activity can then be obtained by measuring the subject's oxygen uptake while performing the task, either with a Douglas bag to collect the expired air or a less restricting spirometer such as the Kofranyi-Michaelis apparatus. Since there are no studies on the energy expenditure of free-living subjects that do not rely, directly or indirectly, on such encumbering apparatus, it is not possible to say whether the use of respirometry introduces errors. Numerous studies of the energy cost of different activities have been made by this procedure. It does not give the net energy cost of each activity above the BMR, but monitors the total rate of energy expended during the period of exercise.

The energy cost is usually expressed per minute rather than per day. The total for 24 hours is then calculated according to the time spent on that activity. The energy cost of a standardized form of physical activity is relatively easy to measure, and it is also possible to estimate the variability in energy cost between individuals in performing the same task. It is much more difficult, however, to obtain accurate values for those tasks that combine a variety of movements, some of which demand the use of heavy parts of the body while others involve only small muscle groups without major weight-bearing or movement of the body.

It is also difficult to generalize on the extent to which differences in body weight affect the energy expenditure for a given type of physical activity. A relation between energy cost and body weight is to be expected when the task involves moving the body, but not when it involves work on external objects. Clearly, many tasks will be a mixture of both types. In the absence of data it has been assumed in this chapter that, regardless of body weight, the same multiple of BMR can be used to express the energy cost of each activity. It is clearly desirable that, wherever possible, investigators should make their own measurements.

A number of workers have estimated the energy expenditure of free-living subjects over periods of several days by monitoring and integrating the number of heartbeats. This procedure relies on indirect calorimetry, since for each individual a calibration curve has to be made relating heart rate to oxygen uptake. The main drawback of this method is that at low levels of activity many physiological and psychological factors may affect the heart rate without appreciably affecting energy expenditure.

Time-scale of estimates

For practical purposes the requirement is generally expressed as a daily rate, although the estimates refer to levels of *habitual* energy expenditure. It is recognized that expenditure varies not only from day to day, but also from week to week. The estimates should represent the average requirement over longer periods. The variability in energy requirements within a group can then be attributed to inter-individual variation only and not to intraindividual variation.

The length of the period chosen will vary with the circumstances. It is well known that there can be substantial variations in the energy requirements of the same subject under different circumstances; for example, the demand for physical activity increases during harvesting in developing countries. At this time of the year energy expenditure may exceed intake and, if so, body weight falls. While some weight loss may be tolerated for a short time, it may be necessary to take account of intermittent periods of heavy work in calculating overall energy needs.

Corrections for metabolizable energy

Once the requirements for energy are obtained from measurements of energy expenditure, the dietary intake needed to meet the energy demand must be determined. Intakes of metabolizable energy have to be calculated to allow for the availability of dietary energy from different sources which may present special problems in some cases, for example a diet rich in fibre. The traditional Atwater factors were designed to allow for non-available energy in different foods, but the corrections for unabsorbed carbohydrate are made in different ways in different food tables. In general, the Atwater factors are still the most suitable in the absence of more specific knowledge on the availability of energy in particular foods.

When intakes are being compared with requirements, it is usually preferable to correct for the metabolizable energy of the diet rather than to adjust the estimate of requirement. For example, if a diet provides only 90% of the metabolizable energy predicted from the Atwater factors, for most purposes it is more convenient to correct the estimated energy intake by a 10% reduction, rather than to increase the requirement by 10%. However, for some uses the latter method may be more appropriate.

Adaptation in Energy Requirements

Adaptation to changes in energy intake can affect energy requirements in three ways; by alterations in body size, by metabolic adaptation, and by behavioural adaptation.

Metabolic adaptation

A very substantial adaptation in total energy requirements can occur on submaintenance intakes, but this involves large changes in most if not all of the factors contributing to total energy expenditure. In the classical Minnesota study of Keys et al. normally nourished subjects were able to achieve approximate energy balance after 6 months on half their usual energy intake. There was, however, a profound fall in body weight (part of which was lean tissue), a reduction in physical activity, and some mental changes, as well as metabolic adaptation shown by a fall in the basal metabolic rate. This degree of adaptation was clearly disadvantageous and the problem is to define the range of adaptation in total energy expenditure that can be achieved without any detectable disadvantage.

When there is a substantial fall in energy intake, in addition to the loss of body weight there is a reduction in the BMR, which declines over a 3-week period by up to 15% when expressed per unit of body weight. Thereafter, further falls in BMR are achieved primarily by a progressive loss of active tissue mass. It is uncertain whether this degree of metabolic adaptation can occur in the absence of a reduction in body size, nor is it clear whether, under conditions of slight energy restriction, there is a fall in the BMR which can then be maintained to bring the body back into energy balance.

When subjects of normal weight are overfed experimentally there is an increase in both lean tissue and body fat, but some metabolic adaptation can also occur. Many overfeeding experiments have been undertaken, but there is little information on total energy expenditure under these conditions. Short-term overfeeding studies have generally shown that most of the extra energy is stored and not dissipated as heat. However, it has been repeatedly confirmed that there are modest increases in the BMR with substantial overfeeding. Long-term overfeeding experiments have been claimed to demonstrate remarkable changes in the body's ability to cope with overfeeding. Unfortunately these studies did not include direct measurements of energy expenditure.

Metabolic adaptation in other components of energy expenditure have been sought under conditions of underfeeding and overfeeding. It has been suggested that there is an interaction between the metabolic response to food and exercise. This response may increase with overfeeding and decline during semistarvation, though the effect is small. There is little evidence that exercise performed under fasting conditions in semi-starved or overfed individuals is appreciably different in efficiency. Finally, studies on the specific dynamic action of a standard meal without exercise in semi-starved or overfed individuals do not suggest major changes in their metabolic response, although in children recovering rapidly from malnutrition the metabolism surges after a meal, and this is related to their rate of growth.

The documented changes in metabolism when energy intake is altered suggest, therefore, that with the present state of knowledge the range of metabolic adaptation must be considered to be small. A variety of mechanisms have been suggested to explain such differences as have been found in the efficiency of energy utilization, including changes in plasma thyroxine and tri-iodothyronine concentrations and in the processes of protein turnover, substrate cycling, and perhaps activation of brown adipose tissue metabolism. Evidence on the quantitative importance of each of these factors is scanty.

Behavioural adaptation

A marked reduction in food intake leads to a profound decrease in physical activity. Children in Guatemala were found to decrease their energy expenditure without changing their growth rate when their dietary energy was reduced by 10%. On the other hand, in studies in Mexico, supplementation of the diet of children led to an increase in physical activity and exploratory behaviour. It has also been shown that when the diet of male agricultural workers in Guatemala was supplemented with additional food, there were appreciable increases in their activity at work and in their discretionary activity without any increase in body weight. There was also an improvement in their sense of wellbeing. These findings suggest that limited food intake makes an appreciable difference to the work capacity of a community and that this happens without, in general, alterations in body weight. Similar improvements in subjective wellbeing with very small weight increases have been found in lactating Gambian women when given supplementary food.

This relationship between energy intake and work output deserves serious consideration during the assessment of energy requirements. The need to provide for the energy cost of socially desirable activity in the home and community has already been emphasized.

Principles of Estimating Protein Requirements

Metabolic Background

Even in the steady state, body proteins constantly undergo breakdown and resynthesis. When growth is occurring, not only is there a net deposition of protein, but the rates of both synthesis and breakdown are increased. The principles underlying this process of protein turnover have been described in detail elsewhere.

The rates of turnover vary from tissue to tissue, and the relative contributions of different tissues to total protein turnover change with age and adaptation to various levels of protein intake. The amino acids released by breakdown are reused for protein synthesis. However, because this process of reutilization is not completely efficient, some amino acids being lost by oxidative catabolism, both essential amino acids and a dietary source of nitrogen are needed. The daily turnover of body proteins is, in fact, several-fold greater than the amino acid intake, showing that the reutilization of amino acids is a major contributory factor to the economy of protein metabolism.

This process of recycling, which includes interchange of amino acids between tissues as well as intracellular reutilization, depends on various metabolic and hormonal factors and is influenced by the physiological status of the host. Thus, reutilization of amino acids is highly efficient during rapid catch-up growth and in convalescence from a catabolic episode resulting from injury or infection. The increased efficiency of reutilization under these circumstances results in the improved use of dietary amino acids, giving the dietary protein an apparently improved biological value.This principle extends to the adaptation of protein metabolism under circumstances of restriction or excess of dietary protein or amino acid supply. Adaptation to a submaintenance level of protein intake leads to a diminished turnover of tissue protein and a reduced rate of catabolism of the amino acids liberated by protein breakdown. In this way, within limits, the tissue protein pool can reach a new steady state appropriate to the diminished intake of protein.

Under the experimental conditions of a protein-free diet, protein synthesis and breakdown continue via the reutilization of amino acids. This process becomes very efficient, but a small proportion of amino acids are still catabolized to urinary nitrogenous compounds and there is some nitrogen loss in the faeces. These represent what has been called the obligatory loss.

It has become clear since the 1971 Committee's report that a key question in relation to the assessment of protein requirements is the extent to which people living on low protein intakes can adapt by increasing the efficiency of recycling and reducing the extent to which amino acids "escape" from the system and are catabolized. One objective, therefore, in determining protein requirements is to define the point at which adaptation is exceeded; beyond this point there will be progressive loss of body protein and deterioration of tissue function.

Adaptation to Low Protein Intakes

The protein requirement has two components—that for total nitrogen and that for essential amino acids—so that a diet may be deficient in quantity or quality of protein. Both aspects may, in theory, be affected by adaptive processes, but almost nothing is known about adaptation of essential amino acid requirements.

All healthy individuals are able to adjust total nitrogen (N) excretion to balance their intake over a certain range. For a time, as the lower limit of this range is approached, body N loss exceeds N intake, and there is a reduction in the mass of body protein leading to a new steady state. This is clearly a form of adaptation. At even lower intakes the limits of adaptation are exceeded, and there will be a continued depletion of body protein resulting ultimately in death. Two questions have to be considered:

(*a*) First, in passing from one steady state to another there is a loss of body protein; is this of any functional significance? When an adult man is transferred from a higher than customary N intake to one that is close to the physiological minimum, e.g., from 14 to 4 g N per day, a level of excretion close to a new steady state is reached in about 7 days, and over this period there is a cumulative N loss equivalent to about 1.5% of total body N. In similar studies on children the new steady state was reached more rapidly, with a total loss estimated at 1% of body N. In experimental animals the losses on passing from a customary to a lower protein intake are relatively greater. From such experiments it is known that initially most of the N is lost preferentially from the liver and gut. Later, because of the recycling discussed above, it is the N content of muscle and skin that is mainly reduced.

It seems doubtful whether a loss of 1–2% of body N in man could represent a significant degree of depletion rather than adaptation. Experimentally, the capacity of animals maintained on low protein intakes to respond to stresses of various kinds does not seem to be impaired, except for an increase in perinatal mortality. Systematic evidence on this question is not available in man, except in children who are demonstrably malnourished, because of the difficulty in matching compounding variables other than nutritional state. Nevertheless, it is not justifiable, given the present state of knowledge, to assert that there are no functional differences between steady states at lower and higher protein intakes. For example, albumin synthesis and breakdown rates, albumin pool size, and perhaps plasma concentration, are somewhat lower at low intakes.

(*b*) The second question is whether, on a habitually low protein intake, it is possible to reduce the lower limit of the range of protein intake at which N equilibrium can normally be maintained. Obligatory urinary N losses in subjects in different countries, presumably with different habitual protein intakes, are remarkably uniform. Thus, the way to reduce the minimum requirement for nitrogen balance is probably to increase the efficiency of utilization of amino acids, i.e., an increase in the recapture of amino acids for protein synthesis, as discussed above, and a decrease in amino acid oxidation.

Although it has been shown in both animals and man that enzymes of amino acid metabolism adapt to changing levels of protein intake, there is no evidence that their activity can be reduced to zero, allowing for 100% reutilization of amino acids. It is possible that some essential amino acids may be more efficiently conserved than others(e.g., lysine). From these considerations it is evident that at the maintenance level the limiting amino acid will be the one that is least efficiently conserved.

The question has been raised as to how far recycling of urea through the gut could contribute to amino acid economy. The ammonia liberated by hydrolysis in the colon is available for the formation of nonessential amino acids. However, this process cannot result in any *net* increase in protein synthesis unless there is a parallel increase in the availability of the essential amino acids or their carbon skeletons—a situation that is unlikely to exist under natural conditions.

Given the evidence currently available, it must be concluded that there is probably only limited scope for metabolic adaptation to N intakes below the physiological minimum for N balance found in subjects with "normal" intakes. The important question that remains is whether the degree of adaptation that does occur, consistent with the maintenance of nitrogen equilibrium, represents a metabolic adjustment without functional significance, or whether it is detrimental to health and long-term survival. The answers to these questions will obviously influence the nutrition of populations subsisting on diets providing low levels of protein. They are also relevant to the determination of protein requirements in healthy, well-nourished individuals.

At the other end of the scale, it is necessary to define the limits of successful adaptation to high protein intakes. It would not be justified to assume that high intakes are automatically optimal. It is known that excessive protein intakes are accompanied by modest elevations in blood urea nitrogen, which facilitates urea excretion, and by an increase in urinary calcium content. Low-birth-weight infants fed very high levels of protein (*5–6*g/kg per day) have in some instances experienced a reduction in growth rate, urine abnormalities, lethargy, and fever, and some studies suggest an impairment in neurological development.

Relationships between Energy and Protein Requirements

The processes of protein synthesis and possibly of breakdown (turnover) require sources of dietary energy and are thus sensitive to energy deprivation. Consequently, the energy balance of the body becomes an important factor in determining nitrogen balance and influences the utilization of dietary protein.

The magnitude of the basal energy needs and of the total amount of protein turned over in a day are both related to active tissue mass. Moreover, in young animals and growing children both rates per unit of active mass are increased compared with those observed in adults. Nevertheless, it has not proved possible to establish a constant numerical relationship, covering all age ranges, between BMR and either protein requirement or obligatory nitrogen loss, although such a relationship has been assumed by previous committees.

There are, however, other ways in which the interactions between energy and protein metabolism are important in relation to protein requirements. It has been known for some time that the utilization of dietary protein is influenced by energy intake and notably by energy balance. It has been demonstrated that, at any given level of dietary protein, addition of energy improves N balance until the response reaches a plateau, which represents the limitations imposed by the dietary protein level. This effect of energy balance can be extended further by raising the protein intake. Studies on animals and on man suggest that increasing the plane of energy intake enhances protein synthesis and reduces amino acid oxidation.

The influence of energy balance on N balance extends from suboptimal up to excess levels of energy intake, so that any change in energy intake above or below the subject's needs is likely to influence his N balance, the effect being of the order of 1–2 mg of N retained per $kcal_{th}$ added (0.24–0.48 mg of N per kJ). This has important implications for the determination of protein requirements when N balance is used as the criterion of adequacy. In view of the difficulty of determining the energy needs of individual experimental subjects, this effect of energy intake must be carefully considered when assessing estimates of protein requirements obtained by the N-balance method.

A less well defined relationship appears to exist between protein intake and the efficiency of utilization of dietary energy. A limited number of studies suggest that changes in intake or utilization of protein produce changes in the rates of weight gain of children and adults under isoenergetic conditions of intake and expenditure. This may be an aspect of the general principle that the improvement in N balance caused by adding energy to the diet can be inhibited when the protein content of the diet is too low.

Requirements for Total Nitrogen

In previous reports a factorial method was used as the basis for predicting the protein requirements of various age groups. This method involved measuring obligatory nitrogen losses (i.e., the amount of nitrogen present in urine, faeces, sweat, etc.) when the diet consumed contained no protein but was otherwise adequate. The requirement for dietary protein was considered to be the amount needed to replace this loss, after adjustments for the inefficiency of dietary protein utilization and the quality of the dietary protein based on its amino acid pattern. For children and pregnant and lactating women, an additional amount of protein (required to support tissue growth and milk formation) was incorporated into this factorial estimate of requirements.

The method adopted by the 1971 Committee has not always been clearly understood. It may be useful to retrace the steps of that Committee's argument and to fill in some gaps where the basis on which estimates were determined is not entirely clear.

Direct measurements on young adult males show that the sum of obligatory losses in urine and faeces is approximately 49 mg of nitrogen per kg of body weight. This figure was derived from the literature and from other studies available to the Committee, which had not been published at the time it met. To this figure was added 5 mg of N/kg to allow for miscellaneous unmeasured losses (sweat, etc.). In order to extrapolate these data to other age and sex groups for which direct results were not available, the Committee, following the example of its predecessor, made use of the general relationship that has been observed in animals of different species between obligatory or endogenous nitrogen loss and basal metabolic rate (BMR).

The BMR of adult males was estimated as 25–27 $kcal_{th}$/kg per day (105–113 kJ/kg per day), so that the obligatory loss was taken as approximately 2 mg of N/basal $kcal_{th}$ (0.48 mg of N/basal kJ). The Committee used this value to estimate the obligatory losses in other groups. Thus, for adult women, the estimate of obligatory loss was reduced by about 10% on the basis of the known sex differences in the BMR of adults.

Since 1971 additional studies have been published on obligatory losses in 2-year-old children, young men, and elderly men and women. There is no further information about losses in older children and adolescents.

The agreement among the values for urinary loss per kg of adult body weight is remarkable, particularly when one considers the difficulty in defining the exact period of time needed for the urinary loss to reach a stable level. The length of the study will probably have less effect on the faecal loss, yet it is the faecal loss that is more variable. It was higher in Indians and Nigerians, presumably because of the nature of their diet.

From evidence available to the 1971 Committee it was apparent that the ratio of obligatory losses to BMR might be significantly less in infants and young children than in adults. The new data from preschool children confirm this difference.

The 1971 Committee was the first to consider the important question of whether the obligatory nitrogen loss can be used to predict the amount of dietary protein needed to meet the minimum physiological requirement. Minimal protein needs had previously been expressed in terms of a hypothetical reference protein that could be used with 100% efficiency; i.e., when fed at the level of the obligatory nitrogen excretion there would be no increase in the urinary and faecal nitrogen losses in excess of those found on a protein-free diet. An ideal amino acid pattern was proposed for the hypothetical reference protein that would provide a standard for determining the quality or number of other proteins.

The 1971 Committee examined the results of balance studies in which egg and milk proteins had been fed to infants, children, and adults at levels below or close to the requirement. This was defined as the amount needed to achieve nitrogen balance in adults and adequate retention in children. It was evident that even these high-quality proteins were not utilized with the efficiency previously assumed, on the basis of animal studies and a few studies in man which had involved very low protein intakes. To meet the minimum requirement, the dietary intake of various age groups had to be 25–50% above that expected from the obligatory losses plus a growth increment. It was proposed that the average requirement for egg and milk protein should be taken as 30% greater than that given by the factorial method for all ages. Re-examination of the data now suggests that the addition should have been of the order of 45%.

The present Consultation has adopted a modified approach, based on that of its predecessor. It was useful to establish the constancy of the obligatory loss. However, once it became clear that N balance could not be achieved simply by replacing the obligatory loss, even with the best quality proteins, the magnitude of that loss becomes of little relevance. The important variable is the efficiency of utilization of dietary protein at the maintenance level. The starting point in determining the present estimates is therefore the direct measurement of the N needed for zero balance in short-term or long-term studies.

Most of the nitrogen balance studies that have been made in recent years have been either on young adult men or on young children. These provide two relatively fixed points, so that the protein requirements of other age groups can be obtained by interpolation, with an allowance, where appropriate, for the growth increment. The BMR was not found to be a satisfactory basis for estimating either the total protein requirement or the maintenance requirement.

Since all the estimates of protein requirements are obtained either directly or indirectly from measurements of N balance, the limitations of balance studies need to be considered in some detail.

Principles of Nitrogen Balance

The nitrogen-balance technique involves the determination of the difference between the intake of nitrogen and the amount excreted in urine, faeces, and sweat, together with minor losses by other routes. In most experiments only the nitrogen content of the diet, urine, and faeces has been directly measured. Allowances are made for losses by other routes on the basis of a limited number of published studies. Thus, to allow for these other losses, any estimate of nitrogen balance limited to measurements of diet and excreta in non-growing individuals must be positive if overall body N maintenance is to be achieved.

In using this method to predict protein requirements, the usual procedure is to feed a series of different levels of dietary protein. The requirement is estimated by extrapolating or interpolating the N-balance data to the zero balance point (N equilibrium) for adults or for adequate growth (positive balance) in children. In early studies the levels fed often included one diet period without protein and other levels of intake far below the requirement. However, from studies on experimental animals and on man, it is known that the N-balance response is *not linear throughout the entire submaintenance range*; the slope decreases considerably as intakes producing zero balance are approached and slightly exceeded.

Accordingly, recent studies have attempted to assess requirements by using several levels of intake that encompass the expected range of requirements. This is one of the reasons why most estimates of requirements based on contemporary studies are higher than those based on data reported in the past. In addition, other variations in experimental design contribute to the differences, such as the level of dietary energy intake and physical activity. In the earlier studies, energy intake was intentionally increased to ensure weight maintenance at low levels of protein intake. However, it is known that this results in more positive (less negative) N balances and therefore lowers the apparent protein requirement.

A second problem is the magnitude of the N losses other than from urine and faeces, the most important of which is via the skin. During heavy work in hot climates appreciable amounts of nitrogen are lost in sweat, mainly as urea, although the losses are lower in those who are acclimatized. There is some evidence of a compensatory reduction in urinary urea output. The N content of sweat is related to blood urea and both decrease with a low protein intake. Other forms of loss, e.g., in hair clippings and menstrual flow, have been measured in detail in some studies.

In the conventional balance study, these losses are not determined and an estimated allowance is made for them. On the basis of the available evidence this allowance has been set at 8 mg of N/kg per day for adults and 10 mg/kg for children up to the age of 12 years. It is unlikely that a single figure will be applicable under all conditions, but there is no realistic alternative to using this method of correction.

A further important consideration is the length of time needed to achieve a steady state at given levels of protein intake. Because adjustments in urinary N excretion do not occur immediately following a change in N intake, it is necessary to allow an adequate period of time for adjustment of N output to the new N-intake level. Most recent experiments concerned with the determination of N requirements have involved diet periods of 1, 2, or 3 weeks at each intake level. This approach characterizes the so-called "shortterm" N balance determinations of protein requirements in man.

The experimental design may be criticized on the ground that the time necessary to achieve a new steady state resulting in zero N balance may in fact be longer than that allowed for in the shortterm studies. If this is so, this method will result in overestimation of the requirements. Data from studies on obligatory N losses show that there is initially a sharp drop in urinary N followed by a long period of relatively stable but slowly declining excretion. The major adjustment appears to be complete by days 5–7 in most adults over a range of age and sex categories and somewhat sooner in children. Because of day-to-day variation in urinary N excretion it is difficult to make accurate measurements of the subsequent slope or to prove that it is significantly different from zero. Most adult subjects receiving protein intakes close to their maintenance needs have not shown statistically significant differences in urinary N output when days 5–10 are compared with days 10–15. However, in a long-term study of men with a low protein intake (0.36 g of protein/kg), it was shown that subjects required from 8 to 28 days to reach a new steady state. From the little evidence available, it appears that when subjects are fed a fixed protein intake over a long period of time, some do and some do not show a slow drift towards a lower rate of urinary N excretion. Data from a very long-term study of men fed 6.5 g of N/day (<" 0.64 g of protein/kg) show that urinary N continued to fall for at least 90 days and perhaps much longer; the rate of decline after the first 2 weeks was about 0.01 g of N/day, which would be statistically undetectable in short-term studies.

It has also been claimed that autocorrelation between day-to-day rates of urinary N excretion will bias the interpretation of balance measurements. Statistically significant autocorrelation has been found in some studies but not in others.

Because of these limitations of short-term balances, longer-term studies should provide a better basis for determining protein requirements. In principle they permit the measurement of other variables that respond more slowly to dietary inadequacy, such as alterations in lean body mass or growth rate in children. In the few long-term studies that have been reported to date the usefulness of various biochemical and other measurements has been explored, e.g., serum albumin and total body potassium (K), but no sensitive and reliable marker has been found. There is undoubtedly a need for other criteria of protein nutritional status as a firm basis for estimating requirements.

Clearly, it would be very difficult to carry out a study in which subjects received a range of different intakes, remaining on each one for several months rather than several weeks. Published long-term balances have so far been conducted at only one level of protein intake. However, longer-term studies under somewhat more realistic conditions, in which adequate growth and positive N balance in children and N equilibrium in adults are maintained, provide the best evidence that the level fed is adequate, whether or not it is a minimum figure.

The preceding discussion assumes that the aim of both short-term and long-term balances is to find the minimum protein intake that will maintain the *status quo*in terms of mass of body protein. This aim allows for only minimal changes in protein mass as the body moves from one steady state to another. Quite large losses of body protein can be supported without loss of life, but the degree of loss that is compatible with optimal function remains unknown.

Requirements for Essential Amino Acids

The requirements for essential amino acids have been assessed by nitrogen balance in adults, starting with the classical work of Rose, by determining the amounts needed for normal growth and N balance in infants and children, and for infants, by comparison with observed intakes of good quality proteins. The report of the 1971 Committee contains estimates of the amino acid requirements of infants, older children, and adults. Since then, information has become available on the amino acid requirements of preschool children aged 2–4 years, and the requirements of infants and older children have been reassessed.

Infants. The 1971 Committee estimated the amino acid requirements of infants by using the intakes of cow's milk formulae or breast milk that supported satisfactory growth. With the exception of the requirement for tryptophan, the values estimated from the amounts of protein consumed in formulae were lower than those determined by Holt & Snyderman from N balances and the growth of infants given amino acid mixtures.

Snyderman et al. were also able to maintain satisfactory rates of growth in a small group of infants who consumed a diluted cow's milk formula to which glycine and urea had been added. These infants were consuming amino acids in amounts slightly below those estimated to be adequate on the basis of the studies of Fomon et al..

Preschool children. Data on the amino acid requirements of preschool children are now available. An experimental design similar to that of Holt & Snyderman was used. The children received diets consisting of 0.3 g of cow's milk protein/kg per day plus an amino acid mixture in proportions and amounts equal to 0.9 g milk-protein/kg per day. The diets provided 100 $kcal_{th}$/ kg per day (418.4 kJ/kg per day) with proper vitamin and mineral supplements. The single essential amino acid under study was partially replaced in the diet by glycine at five different levels. Nitrogen balance was calculated with an allowance of 8 mg of N/kg per day for integumental losses. It was assumed that a retention of 16 mg of N/kg per day (i.e., 100 mg of protein or about 0.5 g of lean tissue gain/kg per day) allowed for normal growth, and results were validated by N-balance studies conducted on children fed milk or soy protein to assess protein needs.

Older children. The amino acid requirements proposed by the 1971 Committee for boys between 10 and 12 years were based on studies in Japan. These values represented the lowest amounts of amino acid needed to bring all subjects into positive N balance. The data have been re-examined by a committee in the USA, which derived estimates of requirements from the information provided on individual subjects. The 1971 Committee's figures are consistently higher than those calculated by the committee in the USA. There are no other studies of amino acid requirements of school-age children, but judging from measurements of N balance in girls of 7–9 years fed two levels of protein, the need for sulfur-containing amino acids is greater than 13 mg/kg and is met by 30 mg/kg. These discrepancies emphasize the limited and unsatisfactory state of knowledge concerning the amino acid requirements of children.

Adults. In the studies on women a positive balance was not necessarily attained; the authors accepted a balance of 0 ± 5% of intake as the criterion of adequacy. The data for the two sexes are not entirely consistent, but whether the differences are attributable to biological differences in the requirements for individual amino acids or to differences in methodology is not known. The requirement for sulfur-containing amino acids has been confirmed by subsequent N-balance studies. The report of the 1971 Committee does not include a value for histidine. Evidence is now accumulating that histidine is essential even for adults, and that this requirement may be between 8 and 12 mg/kg. For other essential amino acids, present information is insufficient to provide more precise figures for adult requirements.

The adult requirement for essential amino acids falls more sharply from infancy than does the total protein requirement. Thus, the proportion of total amino acids (T) that must be supplied as essential amino acids (E) (the E/T ratio) falls with age. This implies that an evaluation of dietary protein quality based on the amino acid requirements or the E/T ratio for infants or young children may underestimate the effectiveness of that protein in meeting the requirements of older children and adults.

General Comments on Methods of Assessing Protein Requirements

In determining protein requirements for maintenance or growth, the present Consultation has relied on direct N-balance studies, although they are not without problems, rather than on the earlier factorial method. The Consultation believes that evidence of N balance in long-term studies constitutes the most acceptable direct evidence. Regrettably, long-term N-balance data are lacking, and the few studies that have been done must be supplemented by results for short-term N balance.

None of the current evidence is entirely satisfactory because there is no method available for the independent validation of an optimal state of protein nutrition. The functional significance of larger and smaller total-body N pools and faster or slower protein turnover rates is unknown. Most biochemical markers (plasma retinolbinding protein, albumin, etc.) are either unchanged even after relatively long periods (30 days or more) of negative N balance or are not readily interpretable (e.g., enzyme changes). There are no functional indicators that can usefully be applied in experimental situations to detect protein inadequacy before clinically detectable changes occur. This area urgently requires further research.

In the final analysis, one would wish to set protein allowances in accordance with such characteristics as health, growth, development, and longevity. This was, in fact, the approach used by our predecessors at the end of the nineteenth and beginning of the twentieth centuries—Voit, Atwater, Benedict, and Cathcart. The majority view, with Chittenden dissenting, appears to have been that protein intakes well in excess of physiologically determined requirements were associated with active and healthy lives.

However, the evidence on which these observations were based is of limited value. Firstly, most habitual diets derive 10–14% of their energy from protein. Thus, when energy intake rises, so does protein intake and also intakes of many of the nutrients associated with protein in foods, such as the B-group vitamins and trace elements. Secondly, it is obvious that many environmental factors will influence any selected measure of health. Populations that characteristically have higher levels of protein intake tend to live under healthier conditions, whereas those with habitually lower intakes are much more likely to be exposed to parasitic and infectious disease. These confounding factors make it extremely difficult to attempt to draw causal relationships. Thirdly, there are many different measures of health and wellbeing; the criteria are therefore complex and cannot easily be used to set physiological requirements for protein.

The Protein-Energy Ratio as a Measure of Dietary Quality

Nutrient-to-energy ratios have attracted wide interest as indices of dietary quality. They have been used in establishing standards of suitable quality (e.g., in the design of diets or in nutrition labelling). They have also been used as indices to be applied when considering whether the quality or the quantity of the diet is likely to be more limiting in particular situations (as in considering whether there may be greater benefit from specific nutrient intervention than from improvement in total food intake). In either mode of application, the underlying question is: "If an individual (or population of individuals) consumes this diet in amounts that will satisfy energy needs, will the concentration of nutrients also be high enough to meet his nutrient needs?"

Platt et al. were largely responsible for the introduction of the ratio of protein energy to total energy (PE ratio) as a convenient and useful descriptor of one aspect of dietary quality in human nutrition. To take into account both the quality and concentration of the protein, they introduced the concept of net dietary protein calories as a percentage of total calories (NDPCals

%). In their work it was applied to the assessment of the adequacy of intakes. For this purpose, the reference PE ratio was calculated as the simple ratio of protein requirements (expressed as equivalent energy) to energy requirements.

In the years that followed, the use of the PE ratio became the topic of much confusion and controversy. It now seems clear that it is a useful concept within limited ranges of application but that, since a reference PE ratio intended to be used as a standard of adequacy can be calculated in different ways, the proposed application of any derived reference ratio must be specified clearly and the limitations to its interpretation must be appreciated.

The calculation of PE ratios which describe known diets is straightforward and noncontroversial. Confusion about the PE ratio relates to the reference criteria to be applied to these calculations and has arisen from three main sources:

a. inappropriate calculation of the reference criterion of a safe or adequate PE ratio as the simple ratio of the recommended or safe level of protein intake (usually, average requirement + 2 SD) to the energy requirement (usually average requirement);
b. failure to recognize the difference between a reference PE ratio applicable to an individual diet (observed intake of an individual) and one applicable to the average diet of a group of population (all individuals not eating the same diet); and
c. failure to recognize that although energy requirements change with activity and life-style protein requirements do not, and hence the criterion of an adequate PE ratio is affected by the physical activity and life-style (actual or desired) of the individual(s).

This last point emphasizes the fact that reference PE ratios are *situation-specific*. Unless such ratios are empirically similar in all situations of interest (a hypothesis that can be examined through appropriate calculations), there can be no universal criterion of a suitable PE ratio.

Derivation of reference PE ratios

Ratios applicable to a particular diet

In considering the nutritional suitability of the diet consumed by an individual, the natural reference standard is the ratio of that individual's protein and energy requirements. As is pointed out in this chapter, the true requirement of a random individual is not known. It can be described only in probability terms as part of the distribution of requirements in a class of similar individuals. It is logical, therefore, that the appropriate PE ratio for an individual should be described as part of the distribution of possible satisfactory ratios of protein to energy requirements among similar individuals. This distribution would include all possible situations in which protein needs would be met when enough of the diet was ingested to meet energy needs. It would take into account the situation in which a random individual's protein needs were high and his energy needs were low as well as when another individual's protein needs were low and energy needs were high. It would consider the likelihood of these situations in any class of individuals who are similar in terms of the variables used to classify individuals for the description of requirements. From the distribution of possible PE ratios, probability statements may be derived concerning the probable adequacy or inadequacy of any particular PE ratio for an individual of the specified type (age, sex, activity, etc.).

Among individuals consuming self-selected diets, it must be expected that the PE ratio of ingested food varies from one person to the next. The average PE ratio for the group does not describe the diet consumed by all members of it.

Studies in North America and Guatemala suggest that the coefficient of variation of the PE ratio of ingested diets among comparable individuals may be some 10–15%. The approach to the development of a reference PE ratio refers to the ratio that should characterize a particular diet, i.e., the one actually ingested by a particular individual. Clearly it is inappropriate to use this as a reference ratio in judging the mean of the distribution of ratios among self-selected diets.

One approach would be to apply the concepts described above to the PE ratio of the diet for each individual and then to aggregate the probability (or risk) statements for the whole group. Another approach would be to derive estimates of average requirements for energy and protein, the variabilities of each, and correlations between them, at the level of the group or population and then to apply the concept and approach developed above with the population as the unit of observation rather than the individual.

Factors that Affect the PE Ratio

Impact of energy requirement on the reference PE ratio

If reference PE ratios are calculated as described above for individuals or classes of individuals, based on the requirement estimates presented in this chapter, it will be found that the requisite PE ratio rises with increasing age after infancy and early childhood. The reference ratio for the older adult (60+ years) is likely to be much higher than that for the preschool child. Although this may appear to contradict commonly held notions about which age groups are more vulnerable to "diet quality", it is a perfectly logical conclusion from the requirement estimates presented in this chapter. The reference PE ratio does not rise because of a progressive increase in protein need per unit of body weight with increasing age; after early childhood there is relatively little change. Rather, the ratio rises because with increasing age there is a progressive fall in estimated energy needs in proportion to body size. The denominator of the ratio falls in relation to the numerator.

A similar phenomenon is observed in young adults as physical activity and hence energy requirements change. For very inactive individuals the reference PE ratio will be quite high; as activity increases, the ratio will fall. Accordingly, it should be recognized that the derived reference PE ratio will be very sensitive to measured or assumed levels of physical activity and hence to particular social situations.

In the present report it has been pointed out that energy requirements may be estimated in terms of the existing body size and composition and existing patterns of activity (the *status quo* approach) or they may be estimated in terms of "desirable" parameters of body size and activity (the normative approach). It is logical to assume that for any stable population, there will have been a variety of accommodations or adaptations that have established an equilibrium between energy intake and expenditure. Thus, it also seems logical to assume that existing energy intakes are adequate or nearly adequate, in most situations, to meet existing energy needs. It is when a normative approach to growth rate, body size in adults, and activity profiles for all ages is introduced that important differences between energy intake and energy requirement appear.

From the above it follows also that in calculating PE ratios there are two approaches that may be followed. The *status quo* approach would accept existing energy intake as an estimator of existing energy requirement and use this in the calculation of PE ratio. The normative approach would accept the energy requirement that would be associated with a desired state of growth rate, body size, and activity and use this in calculating the PE ratio. In so far as growth rate and body size descriptors differ in these two approaches, estimates of protein requirement as well as energy requirement would be affected. However, for differences in activity profiles, only the energy requirement estimates would be influenced.

These two approaches can be expected to yield quite different estimates of appropriate PE ratios. The *status quo* approach may yield ratios some 15–20%*higher* than the normative approach.

In the past decade or so, since the publication of the report of the Joint FAO/WHO *Ad Hoc* Expert Committee on Energy and Protein Requirements, it has been argued by many that both energy and protein intakes are low in comparison to requirement estimates for low income populations in developing countries. However, it also has been argued that if actions are taken to increase energy intakes to approximate requirement estimates, the accompanying increase in protein intake would meet or surpass requirement estimates (i.e., the PE ratio of existing diets is adequate but the total level of food intake is low). If one accepts the *status quo* approach to energy intake and requirement, the argument of relative adequacy may be fallacious. *If* existing energy intakes and the associated body sizes and activity profiles are deemed acceptable, then it is appropriate to assess protein intake without major regard to energy (or to calculate PE ratios using existing energy intakes as the estimate of energy requirement). This could lead to a different conclusion about the probable adequacy of existing diets, at least for some population groups.

Although recognizing the practical importance of these considerations, the Consultation did not attempt an analysis of dietary data to explore the implications. Rather, this discussion is presented to emphasize the need for caution in the calculation, application, and interpretation of PE ratios. The Consultation also emphasized the need to consider very carefully the goals of interventions (in terms of growth rates, body sizes, and activity profiles) in any consideration of the suitability of nutrient: energy ratios.

Catch-up growth as a special situation

During rapid growth there is a change in both energy and protein requirements per unit body size. Because these changes differ in relative magnitude, the appropriate reference PE ratio will change. The method used to calculate the reference PE ratio does not change, but the requirement estimates will differ. The important practical point is that while reference PE ratios may be expected to rise with increasing assumed growth rates, there may be practical external limitations on the rate of catch-up growth that can be achieved by children in the community; there would be little merit in establishing reference PE ratios for growth rates that cannot be attained. Similarly, if the quantity of food offered to a child limits the rate of catch-up growth, a higher concentration of protein in the food may have no real value. A careful analysis of the situation is needed before reference PE ratios for catch-up growth can be calculated and interpreted meaningfully.

Adjustment for digestibility and amino acid composition of dietary protein

To be useful the PE ratio must take into account the digestibility and amino acid composition of dietary protein. Reference PE ratios could be adjusted upward to allow for the digestibility and amino acid composition of diets ingested by particular age groups. This was the method adopted in the past. However, in many applications a more useful approach would be to adjust the observed protein intake to a value that represents the intake of utilizable protein. The reference PE ratio would then continue to be expressed in terms of protein equivalent to egg or milk protein and the dietary protein would be adjusted to this base. This is analogous to the approach of Platt et al. in their use of the term NDPCals %, although the methods of calculating the reference PE ratio and adjusting for protein quality are different in detail.

Other measures of dietary quality

This section has discussed only two parameters of the nutritional quality of diets, energy and protein, combined in the form of protein : energy ratios as an index of diet quality. It must be recognized that the concentration of all other necessary nutrients should be considered when assessing the adequacy of a particular diet or establishing standards of nutritional quality for foods and diets. It is potentially dangerous to consider energy and protein alone.

There is a special situation where the PE ratio (and other nutrient : energy ratios) of a diet may appear to be adequate, but where it can be predicted that insufficient amounts of the diet will be consumed to satisfy either energy or nutrient needs. This happens when young children are offered a too bulky diet with a low energy density. The volume of food to be consumed may exceed the capacity of the child and hence the PE ratio will be meaningless. The shortfall between the volume of food that a child needs to consume to satisfy requirements and the volume that it is capable of consuming depends on both the energy density of the food mixture and the pattern of the meals provided. The presumed limitation is the physical capacity of the gastrointestinal system to accept food offered at particular times. In this situation the constraints on food intake may be reduced either by changes in the feeding pattern or the energy density. It has been suggested that adding oil to increase the energy density would be a practical solution in many cases. If the PE ratio of the original diet were already low, adding oil would be expected to lower the ratio even further. Comparison of the resultant PE ratio with reference PE ratios might in some situations indicate that the addition was inadvisable. This would be an example of an appropriate use of PE ratios in assessing the quality of diets. This example also illustrates that much more than the PE ratio must be taken into account when assessing the quality of diets in the community.

References

Eveleth, P.B. & Tanner, J.M. 1976. *Worldwide variation in human growth.* Cambridge, Cambridge University Press.

Ferro-Luzzi, A. et al. 1979. Nutrition, environment and physical performance of preschool children in Italy. In: Somogyi, C. & de Wijn, J.F., ed. *Nutritional aspects of physical performance.* Basel, Karger, pp. 85–106.

Garn, S.M. In: Farmer, F.A., ed. 1978. *Nutrition of the aged.* Alberta, University of Calgary, pp. 73–90.

Paøízková, J. In: Chandra, R.K., ed. 1981. *Critical reviews of tropical medicine*, Vol. 1. New York, Plenum Publ. Corp.

World Health Organization. 1983.*Measuring change in nutritional status.* Geneva, World Health Organization.

6

Role of Dietary Fats and Oils

The role of dietary fats and oils in human nutrition is one of the most important areas of concern and investigation in the field of nutritional science. The findings of these investigations have wide-ranging implications for consumers, health-care providers and nutrition educators, as well as food producers, processors and distributors. New evidence concerning the benefits and risks associated with particular aspects of dietary fat is constantly emerging in both the scientific literature and the popular media. At times, controversies about these findings evolve. Sifting through all the claims and counterclaims, incomplete and incompatible studies, biases and competing interests, for the elements of truth and a prudent course of action is a challenge. Yet, this task is essential because changing views about the effects of dietary fats and oils can profoundly influence the consumption of various foods and, ultimately, health and nutritional status, agricultural production, food processing technologies, food marketing practices and nutrition education.

Composition of Dietary Fat

Fats occur naturally in food and play an important role in nutrition. Fats and oils provide a concentrated source of energy for the body. Fats are used to store energy in the body, insulate body tissues, and transport fat soluble vitamins through the blood. They also play in important role in food preparation by enhancing food flavour, adding mouth-feel, making baked products tender, and conducting heat during cooking. Fats and oils are made up of basic units called fatty acids.

Dietary fat includes all of the lipids in plant and animal tissue which are eaten as food. The most common fats (solid) or oils (liquid) are a mixture of triacylglycerols (triglyceride) with minor amounts of other lipids. The fatty acids present in various lipid molecules are the moieties of great nutritional interest.

Fatty Acids

The most abundant fatty acids have straight-chains of an even number of carbon atoms. There is a wide spectrum of chain-lengths, ranging from a four-carbon fatty acid in milk to thirty-carbon fatty acids in some fish oils. Frequently, the fatty acids have eighteen carbons. Double bonds along the carbon chain or substituents on it are designated chemically by counting the carboxyl carbon as position 1. Thus, the double bonds in linoleic acid give it the chemical systematic name of 9, 12-octadecadienoic acid.

A short-hand abbreviation for linoleic acid is 18:2 (eighteen carbon atoms: two double bonds). Its last double bond is six carbons from the methyl end, an important feature for some enzymes. This is considered a n-6 or w 6 fatty acid. The common (trivial) name, the chemical systematic name and the short-hand abbreviation for some dietary fatty acids are provided in table 1. The double bonds in the fatty acids are in the cisconfiguration. The first member of the n-6 series of fatty acids is linoleic acid and the first member of the n-3 series is α-linolenic acid.

Table 1: Some dietary fatty acids

Common name Fatty acid family	Systematic name	Abbreviation	
capric	decanoic	10:0	
lauric	dodecanoic	12:0	
myristic	tetradecanoic	14:0	
palmitic	hexadecanoic	16:0	
stearic	octadecanoic	18:0	
arachidic	eicosanoic	20:0	
behenic	docosanoic	22:0	
lignoceric	tetracosanoic	24:0	
palmitoleic	9-hexadecenoic	16:1	n-7
oleic	9-octadecenoic	18:1	n-9

gadoleic	11-eicosaenoic	20: I	n-9
cetoleic	11-docasaenoic	22:1	n-11
erucic	13-docasaenoic	22: 1	n-9
nervonic	15-tetracosaenoic	24:1	n-9
linoleic	9,12-octadecadienoic	18:2	n-6
a -linolenic	9,12,15-octadecatrienoic	18:3	n-3
g -linolenic	6,9,12-octadecatrienoic	18:3	n-6
dihomo-g -linolenic	8,11,14-eicosatrienoic	20:3	n-6
	5,8, 11-eicosatrienoic	20:3	n-9
arachidonic	5,8,11,14-eicosatetraenoic	20:4	n-6
EPA	5,8,11,14,17-eicosapentaenoic	20:5	n-3
adrenic	7,10,13,16-docosatetraenoic	22:4	n-6
	7,10,13,16,19-docosapentaenoic	22:5	n-3
DPA	4,7,10,13,16-docosapentaenoic	22:5	n-6
DHA	4,7,10,13,16,19-docosahexaenoic	22:6	n-3

The n-6 and n-3 polyunsaturated fatty acids have cis double bonds that are interrupted by methylene groups. A double bond may change from a cis to a trans configuration or move to another position along the carbon chain as illustrated in Figure 1. The shape of a trans fatty acid is similar to that of a saturated fatty acid. As a result, trans fatty acids have a higher melting point than their cis isomer. The trans isomer may be regarded as an intermediate between an original cis unsaturated fatty acid and a completely saturated fatty acid.

Figure 1: Structure of cis and trans double bonds

Acylglycerides. The type of fatty acid and the position in which it is esterified to glycerol determine the characteristics of acylglycerides. In addition to glycerides which have three esterified fatty acids, diacylglycerides and monoacylglycerides occur in raw food or food ingredients. There is some specificity in the position occupied by the fatty acids. Animal depot fats tend to have a saturated fatty acid in position 1 and an unsaturated fatty acid in position 2. Fatty acids in position 3 appear to be more randomly distributed, although polyunsaturated fatty acids often accumulate there.

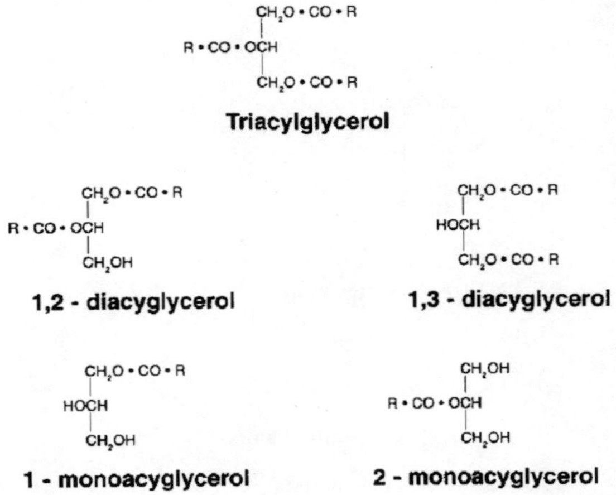

Figure 2: Diagram of acylglycerides

Phospholipids. Phospholipids are components of membranes which occur in foods and extracted oils. The general structure of phosphoglycerides is shown in Figure 3. A saturated fatty acid is usually esterified at position 1 and a polyunsaturated fatty acid at position 2. The polar groups which contain phosphorus and an organic base provide the lipid molecule with a hydrophilic region. In addition to phosphoglycerides, phospholipids include sphingomyelins and cerebroside which are based on sphingosine rather than glycerol. Although phospholipids constitute only a small fraction of total dietary fat, they can be an important source of essential fatty acids.

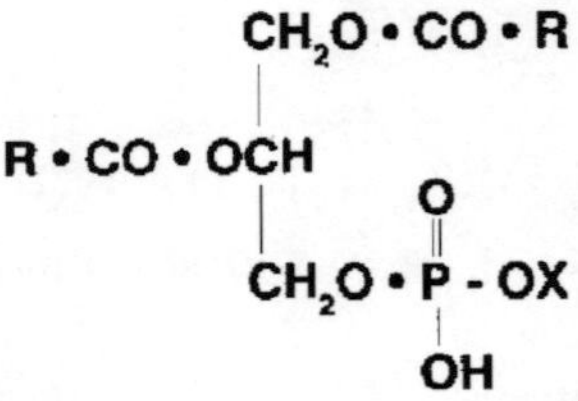

Phospholipid where X is choline, ethanolamine, serine, inositol, glycerol.

Figure 3: Diagram of phospholipids concentrations relative to triacylglycerols.

Nonglyceride components

The growing realisation of the importance of the nonglyceride components, sometimes called "minor constituents", in fats and oils warranted inclusion of this topic in the expert consultation.

Vitamin E: Vitamin E consists of a mixture of lipid-soluble phenols characterised by an aromatic chromanol head and a side chain of 16-carbon atoms. The tocopherols have a saturated hydrocarbon tail, whereas the tocotrienols are the farnesylated analogues having an unsaturated isoprenoid tail. The number and position of the methyl groups on the chromanol ring give rise to the different α–, β–, γ- and β-tocopherol and tocotrienol isomers.

Vegetable oils and the products made from them usually contain large amounts of tocopherol, especially the α, β and γ isomers. In addition, certain vegetable oils, particularly palm oil and rice bran oil, are rich sources of tocotrienols which have weak vitamin E activity but act as antioxidants and provide stability against oxidation.

Carotenoids: Carotenoids are highly unsaturated polyisoprene hydrocarbons that are lipid soluble. Over 75 different carotenoids are known to occur in animal fats and vegetable oils. The most common are α, β and γ carotene, Iycopene, lutein and xanthopylls. The carotenoids and their derivatives are generally responsible for the yellow to deep-red colour of fruits, vegetables, cereals and crude palm oil. Carotenoids are the precursors of vitamin A, with ß-carotene having the highest provitamin A activity.

Vitamins A and D: A traditional source of vitamin A is butterfat. Fish oils are a common source of vitamin D. Margarines, which are fortified with vitamins A and D by law in most countries, also make an important contribution to ensuring adequate intakes of these nutrients.

Other Components

Sterols: Cholesterol is the principal sterol of animal products. The major sterols of plants are β-sitosterol, campesterol and stigmasterol, although several others are known to exist.

Methylsterols and triterpene alcohols: Sterols methylated at the 4-OH position occur in common vegetable oils at concentrations of 0.01 to 0.4 percent with rice bran oil and sesame oil having the highest levels. Corresponding concentrations of triterpene alcohols, comprised of five condensed cyclohexane rings, are 0.01 to 1.2 percent. Rice bran oil is the only one at the upper level.

Squalene: The predominant hydrocarbon in dietary fat is squalene. It is an intermediate in the synthesis of sterol from acetate, and it is found in particularly high quantities in some fish and olive oil. The concentration in most vegetable oils is below 30 mg/100 g.

Oryzanols: Oryzanols are compounds consisting of ferulic acid esterified to a variety of plant sterols and triterpene alcohols. Although large amounts are found in crude rice bran and linseed oils, oryzanols are not widely distributed in other oils.

Digestion and Metabolism

Most dietary fat is supplied in the form of triacylglycerols which must be hydrolysed to fatty acids and monoacylglycerols before they can be absorbed. In children and adults, fat digestion is efficient and is nearly completed in the small intestine. In the newborn, the pancreatic secretion of lipase is low. The digestion of fat in babies is augmented by the lipases secreted from the glands of the tongue (lingual lipase) and a lipase present in human milk. The stomach is part of the process of fat digestion because of its churning action which helps to create an emulsion.

Fat entering the intestine is mixed with bile and is further emulsified. The emulsion is then acted upon by lipases secreted by the pancreas. Pancreatic lipase catalyses the hydrolysis of fatty acids from positions 1 and 3 to yield 2-monoacylglycerols. Phospholipids are hydrolysed by phospholipase A2 and the major products are lysophospholipids and free fatty acids. Cholesterol esters are hydrolysed by pancreatic cholesterol ester hydrolase.

The free fatty acids and monoglycerides are absorbed by the enterocytes of the intestinal wall. In general, fatty acids which have a chain length of less than 14 carbons enter directly into the portal vein system and are transported to the liver. Fatty acids with 14 or more carbons are re-esterified within the enterocyte and enter the circulation via the Iymphatic route as chylomicrons. However, the portal route has been described as an absorptive route for dietary long chain fatty acids as well.

Fat soluble vitamins and cholesterol are delivered directly to the liver as part of the chylomicron remnants. Diseases that impair the secretion of bile, such as biliary obstruction or liver diseases, lead to severe fat malabsorption, as do diseases that influence the secretion of lipase enzymes from the pancreas, such as cystic fibrosis. As a result, medium-chain triglycerides can be better tolerated in individuals with fat malabsorption and these are often used as a source of dietary energy. Complete absorption of lipids from the intestine may be marginally affected by a high amount of fibre in the diet.

Fatty acids are transported in the blood as complexes with albumin or as esterified lipids in lipoproteins. These consist of a core of triacylglycerols and fatty acid esters of cholesterol, and a shell of a single layer of phospholipids interspersed with unesterified cholesterol. Coiled chains of one or more apolipoproteins extend over the surface and, with the amphipatic phospholipids, enable the lipids in the core to be carried in the blood. They also regulate the reaction of the lipid package with specific enzymes or bind the particle to cell surface receptors.

Chylomicrons are lipoprotein particles derived from dietary fat and packaged by the mucosa cells. They enter the blood stream via the Iymph vessels. Lipoprotein lipase located on the interior walls of the capillary blood vessels hydrolyses the triacylglycerols, releasing fatty acids. These enter the adipose tissue where they are stored, and the muscles where they serve as fuel. The remnants of chylomicrons are cleared by the liver within a few hours of the ingestion of a fat-containing meal.

Very-low density lipoproteins (VLDL) are large triacylglcerol-rich particles produced in the liver from endogenous fat, as opposed to chylomicrons which transport exogenous fat. VLDLS are the main carriers of triacylglycerols which are also processed by lipoprotein lipase and supply fatty acids to adipose and muscle tissues.

Low-density lipoproteins (LDL) are the end products of VLDL metabolism. The core consists mainly of cholesterol esters and its surface has only one type of apolipoprotein, apoB. About 60-80 percent of cholesterol in plasma is carried by LDLs. Average LDL values vary among populations because of genetic and environmental factors, however, diet is probably a major determinant of these values.

High-density lipoproteins (HDL) carry 15-40 percent of plasma cholesterol. They are probably formed in the circulation from precursors made in the liver and in the gut. The major apolipoprotein of HDL is apoA-1. In humans, LDL delivers cholesterol to the liver and HDL may transfer cholesterol to other lipoprotein particles. There is evidence that HDL actively protects vessel walls. It is not known whether manipulation of HDL levels by diet affects the development of atherosclerosis.

Lipoprotein(a) or Lp(a) is a complex of LDL with apolipoprotein(a). This apoprotein has a sequence homology with the pro-enzyme plasminogen which is involved with the dissolution of blood clots. The concentration of Lp(a) is largely determined by genetics.

Metabolism

Saturated fatty acids and monounsaturated fatty acids can be biosynthesised from carbohydrates and proteins. Unsaturated fatty acids may be substrates for desaturases and elongases, as shown in the scheme for conversion in the n-9, n-6 and n-3 families of fatty acids.

The first members of each fatty acid family (oleic, linoleic and α-linolenic acid) compete for the same 6-desaturase, with rates of conversion increasing with the number of double bonds. This ratelimiting enzyme is under the control of many dietary and hormonal factors and is believed to be important in the synthesis of 22:6 n-3. Such an effect may explain why high intakes of linoleic acid reduce the level of 22:6 n-3. Similarly, the 5-desaturase is modulated by dietary and hormonal factors.

The C20 and C22 members of the n-6 and n-3 families can inhibit desaturation earlier in the sequence of fatty acid conversion. It appears that the 4-desaturation does not involve another specific desaturase, but an elongation, then 6-desaturase, both microsomal processes, followed by retroconversion by the peroxisomal, oxidation pathway. The desaturation is thus a process based on intracellular cooperation. The nutritional significance of this additional role for 6-desaturase has yet to be evaluated. The activities of the elongases appear to be greater than those of desaturases.

Essential Fatty Acids

The essential fatty acids are those that must be supplied in the diet and they include members of both the n-6 and n-3 series. Under conditions of a dietary deficiency of linoleic acid, the most abundant unsaturated fatty acid in tissue, oleic, is desaturated and elongated to eicosatrienoic, n-9, which is normally found in trace amounts only. The accumulation of this fatty acid is considered to be a marker of essential fatty acid deficiency. When the diet has a low n-3 fatty acid content, compared to the n-6, there is a reduction of 22:6 n-3 associated with a compensatory accumulation of 22:5 n-6 in tissues (Gall), Agradi and Paoletti.

The most abundant product of the n-6 pathway is arachidonic acid found in glycerophospholipids where it is selectively esterified in the 2-position. Diets containing fats of animal origin provide appreciable amounts of preformed arachidonic acid. The efficiency of the n-3 pathway in humans has been open to question. However, human studies on the conversion of deuterated linoleic and α-linolenic acids to their long chain derivatives showed that the efficiency for the conversion of linoleic to arachidonic was 2.3 percent, while the conversion of α-linolenic to its long chain derivatives was 18.5 percent. These conversion values, measured as the accumulation of products in plasma, were obtained in subjects with relatively low intakes of linoleic acid (around 5 percent).

When the diet contained around 9 percent linoleic acid, the conversions were reduced to 1 percent for linoleate and 11 percent for linolenate. These data do indicate that the synthesis of n-3 metabolites from 18:3 occurs in adults. Fish oil, which is rich in eicosapentaenoic acid compared to linseed oil, greatly enhances the concentration of the long-chain derivatives in human blood. High levels of eicosapentaenoic acid and docosapentaenoic acid in blood platelets or other cells are attained only when they are provided as such in the diet. The n-3 fatty acids, like the n-6 fatty acids, are incorporated into the 2-position of membrane phospholipids.

The polyunsaturated fatty acids for membrane phospholipids come from the diet or endogenous sources. The availability is influenced by deacylation-reacylation, particularly at position-2 of phospholipids, the specificity of diacylglycerol kineses in phospholipid synthesis, and remodelling of donor and acceptor phospholipids. In all of these processes, competition exists between the fatty acids of the n-6 and n-3 series.

The essentiality of linoleic acid was demonstrated in studies of rats. When rats were fed a fat-free diet, they developed specific symptoms that were prevented by the provision of linoleic acid. Linoleic acid deficiency, a rare condition in humans, was described in children given fat-free diets. The children developed skin disorders similar to those produced in rats that were cured by the administration of linoleic acid. To prevent the accumulation of ecosatrienoic acid (20:3 n-9), the amount of linoleic acid required in the diet is at least 1 percent of the energy intake. This applies to all the animal species tested.

Intakes between 4 and 10 percent of the energy are currently consumed by different population groups, and these appear to be compatible with optimal health status. Linoleic acid deficiency may develop as a secondary condition in other disorders, such as proteinenergy malnutrition and fat malabsorption, or as a consequence of total parenteral nutrition with inadequate linoleic acid intakes. Under conditions of adequate linoleic acid intake, the ratio of triene to tetraene is below 0.2.

A better assessment of linoleic acid deficiency can be obtained with a profile of all polyunsaturated fatty acids in serum. The n-3 fatty acid, docosahexaenoic acid, is present in high concentrations in the central nervous system, in the cell membranes, and the visual system. The balance between n-6 and n-3 fatty acids in the diet is important because of their competitive nature and their essential and different biological roles. It is suggested that the relative amounts of linoleic and α-linolenic acids in the diet should be below 10:1.

Biological Roles of Fatty Acids

Membrane structure. Unsaturated fatty acids in membrane lipids play an important role in maintaining fluidity. In the skin, linoleic acid plays a specific role where it is linked to some very long chain fatty acids (C30-C34) in the acyl ceramides. These form an intercellular matrix to maintain the epidermal permeability barrier. In membranes, interactions between lipids and proteins may depend upon a specific polyunsaturated fatty acid. This appears to be the case for mammalian rod outer segments which are very rich in docosahexaenoic acid.

Three examples of lipid-protein interactions that control the metabolic functions of membranes were noted: first, the catalytic properties of transporter proteins; second, the activities of enzymes such as the Ca/Mg ATPase of sarcoplasmic reticulum, adenyl cyclase, and 5-nucleotidase which are influenced by the levels of n-6 and n-3 fatty acids in the membrane lipids; and third, enzymes involved in the phosphoinositide cycle. This cycle, associated with the responses of many cells to a range of hormones, neurotransmitters and cell growth factors, gives rise upon activation of a specific phospholipase C to two important second messengers: inositol 1,4,5 trisphosphate and diacylglycerol.

IP3 is responsible for the modulation of cytosolic calcium ions. Diacylglycerol, together with calcium ions and phosphatidyl serine, is involved in the activation of a protein kinase C which phosphorylates a number of intracellular proteins. Since polyphosphoinositides are very rich in arachidonic acid in position 2, diacylglycerol arising from them is also rich in this fatty acid. A lipase acting on diacylglycerol may subsequently release arachidonic acid for conversion to eicosanoids.

The phosphoinositide cycle controls processes in cell division. The modulation of the cycle, therefore, influences the rate of division of cells in the immune response and in tumour growth. Dietary polyunsaturated fatty acids and the proportion of n-6 and n-3 fatty acids appear to affect the phosphoinositide cycle.

Eicosanoid formation. The n-6 and n-3 fatty acids in the membrane phospholipids exert metabolic control through their role as precursors of eicosanoids. These highly active compounds of 20 carbon atoms are released in very small quantities to act rapidly in the immediate environment. After enzymatic degradation, the products from eicosanoids in the urine are an indication of the production by the body. The first step in the biosynthesis of eicosanoids is the release of a 20 carbon polyunsaturated fatty acid from phospholipids by phospholipases, mainly a phospholipase A2, or from diacylglycerol produced in the inositol phosphate cycle.

The eicosanoid cascade consists of hydroxylated derivatives of 20 carbon polyunsaturated fatty acids: (a) cyclic products, generated by a cyclo-oxygenase, which include prostaglandins, prostacyclin and thromboxane; (b) lipoxygenase products such as the 12-lipoxygenase derivatives, especially the products of the 5-lipoxygenase known as leukotrienes; and (c) products of cytochrome P450 activity. Eicosanoids, in general, are very potent, their effects are highly diverse, and the action of different eicosanoids often oppose each other. In addition, the patterns of eicosanoid production are different in various cells and tissues.

Among the most active eicosanoids, thromboxane A2(Tx A2) derived from arachidonic acid is produced by platelets and other cells through the cyclo-oxygenase pathway. This eicosanoid is a platelet proaggregatory and smooth muscle contracting agent, and is rapidly inactivated to thromboxane B2. Prostacyclin (PGI2), produced through the cyclo-oxygenase in cells of vessel walls, is a platelet anti-aggregatory and vasodilating agent. Other products of the cyclo-oxygenase pathway exert various effects on smooth muscle cells, immune competent cells, and so forth.

Among the products of the lipoxygenase pathway, the leukotrienes, produced mainly by the leukocytes, act on vascular parameters (permeability, contractility) and have chemotactic properties. They are involved in the modulation of inflammatory and immune processes. Polyunsaturated fatty acids with 20 carbons and different degrees of unsaturation give rise to eicosanoids with different numbers and patterns of unsaturation, and with somewhat different biological activities. Since arachidonic acid is the major polyunsaturated fatty acid in cells, the 2-series of eicosanoids is the most abundant and generally the most active.

When a 20 carbon polyunsaturated fatty acid with a different number of double bonds, for example, di-homo-γ linolenic, 20:3 n-6, or eicosapentaenoic acid (EPA), 20:5 n-3, is incorporated into cell lipids, eicosanoids of the 1-series or the 3-series, respectively, are produced. These fatty acids also compete with arachidonic acid for the cycloxygenase, and thus reduce the formation of eicosanoids of the 2-series. EPA promotes the formation of eicosanoids of the 3 series and inhibits the formation of eicosanoids of the 2 series.

Effects on other lipid-derived mediators. Hydrolysis of selected cell phospholipids results in the formation of additional biologically active compounds such as the platelet activating agent, PAF. This compound arises from l-alkyl, 2-acyl phosphatidyl choline. PAF is an extremely potent pro-inflammatory agent, and a potent activator of several types of cells in addition to platelets. Although data on the effects of various polyunsaturated fatty acids on the PAF pathway are limited, incorporation of arachidonic acid into cell phospholipids was shown to potentiate PAF production. Different polyunsaturated fatty acids in position 2 of the precursor phospholipids can modify PAF formation, as was shown by the reduced production of PAF by monocytes of subjects receiving n-3 fatty acids. The opposing effects of n-6 and n-3 fatty acids may explain some influences of polyunsaturated fatty acids on the function of certain cells.

Effects on other parameters. The n-3 fatty acids appear to affect various other processes, such as the production of cytokines and other factors. Cytokines are a family of proteins produced and released by cells involved in inflammatory processes and in the regulation of the immune system. These include the interleukins and tumour necrosis factors. The mechanism by which n-3 fatty acids affect cytokine synthesis is not clear, but some studies have shown an effect on mRNA levels, suggesting a pretranslational level of action.

The effects of fatty acids on the expression of genes encoding for enzymes which are involved in lipid metabolism, as well as on the expression of genes involved in cell growth regulation, represent an additional important aspect of the biological roles of polyunsaturated fatty acids. It is apparent that fatty acids can interact with a group of nuclear receptor proteins that bind to certain regions of DNA and thereby, alter transcription of regulatory genes. Much of the literature on the n-3 long-chain fatty acids deals with pharmacological doses or is concerned with lipoproteins and their relationship to coronary heart disease. Additional clinical and metabolic studies should be encouraged.

Oxidation

Those fatty acids that are not used for synthesis of eicosanoids or incorporated into tissues are oxidised for energy. Fatty acids yield energy by beta oxidation in the mitochondria of all cells, except those in the brain and kidney. They enter the mitochondria as specific acyl carnitine derivatives. Saturated short, medium, and long-chain fatty acids undergo the first step of beta-oxidation with different dehydrogenases. The process yields successive acetylCoA molecules which enter the tricarboxylic acid cycle or other metabolic pathways. Acetate is the eventual product from fatty acids with an even number of carbon atoms.

Unsaturated fatty acids require two more enzymatic steps than saturated fatty acids to change cis double bonds to trans and to move them from the a to beta position. Even so, oxidation of unsaturated fatty acids, including linoleic acid is as fast or faster than that of palmitic acid. The initial oxidative reaction is carried out by a different enzyme from that in mitochondria; the fatty acyl CoA enters directly into this organelle. The process does not proceed to the complete production of acetate but a shortened fatty acid is transferred to the mitochondria for complete oxidation.

Long chain (>20C) fatty acids are preferentially oxidised by peroxisomes; also fatty acids with less than 14C are oxidised by this system. Peroxisomal oxidation is energetically less efficient than mitochondrial oxidation and yields more heat. This type of oxidation can be induced by diets which are high in fat as well as by a variety of xenobiotics.

Availability of Edible Fats and Oils

Consumers are often attracted to foods which have textures and flavours derived from fat. Although there are differences according to region, season and food habits, consumers generally increase the proportion of fat in their diet as their incomes rise. The increase in the quantity and change in the quality of fats and oils in the diet have important consequences for nutrition.

In 1990, the amount of total dietary fat that was available worldwide was estimated to be 68 grams per caput per day. However, this average figure does not reveal the large disparities among geographic regions. While in Asia and Africa the amount of total fat available was less than 50 grams per caput per day, in South America it was 74 grams per caput per day. In the former USSR, the amount of total fat available was 107 grams per caput per day. In North and Central America, the total amount of fat available was 126 grams per caput per day while in Europe it was 143 grams per caput per day. Finally, 121 grams per caput per day were available in Oceania.

When countries are categorised by level of economic development, the differences in the availability of total fat among groups appear more distinctly. In developed countries, the daily per caput availability of fat was 128 grams while in developing countries it did not exceed 49 grams. In Africa, the availability of fats and oils is low, while in North America, the level of fat available is high. Within these two economic groups, there are large differences in total fat availability among regions and individual countries. Countries such as Rwanda, Cambodia and Bangladesh have less than 20 grams per caput per day while Ireland, Denmark, Luxembourg and Belgium have more than 170 grams per caput per day.

The Fat Energy Ratio (FER) is the proportion of dietary energy derived from total fat. Among individual countries, the average FER ranges from 7 to 46 percent. Survey data on food consumption tend to confirm these trends and patterns of use of dietary fats and in some cases, demonstrate even more extreme situations. In Viet Nam, for instance, the average consumption of fats and oils was found to be very low. The FER calculated on the basis of consumption surveys was 6 percent while it was 11 percent according to the Food Balance Sheets. In some areas of Viet Nam (such as the "Highlands") data from these surveys indicate that the consumption of fat and the FER are even lower. On the other hand, the MONICA project in Germany indicated that the FER for men was more than 50 percent, if alcohol is excluded from the energy breakdown.

FAO Food Balance Sheets show that the availability of fat for human consumption has increased steadily in both developed and developing countries. From 1961 to 1990, global availability of fat increased from 49 to 68 grams per caput per day. In developed countries, the amount of fat available grew from 93 to 128 grams while it increased from 28 to 49 grams in developing countries. Even though the increase in fat availability was twice as high in developing countries as in developed ones, there was still a large difference in the availability of fat according to levels of economic development.

Within developing regions, the rates of increase have varied. In 1961, the fat availability in the Far East was lower than in Africa. Since then, the availability of fat has increased sharply and it is now higher than in Africa. The availability of fat was high in Oceania in 1961 and the increase has been minor. Fat availability was low in Africa 30 years ago and the increase in availability has been slow.

Income. The first factor which explains these changes is income, the basic indicator of economic development. The availability of animal and vegetable fats is closely linked to income. There was a steady increase in fat availability in countries with incomes ranging from US$150 to US$350 per year. For those countries with yearly per caput incomes between US$350 and US$7 000, there was a sharp increase in the amount of total fat available, while the availability of fat remained at the same level for countries with individual incomes above US$7 000.

In countries with per caput incomes under US$ 7 000, consumption of both vegetable and animal fats increased at similar rates. Beyond US$ 900, the availability of animal fats increased rapidly. The availability of vegetable fats actually declined when per caput incomes exceeded US$7 000. Food consumption surveys confirm this trend especially in countries such as Bangladesh or Brazil Instituto Brasileiro de Geografia e Estadistica where GNP/caput is low. In these countries, poverty is the main factor limiting fat consumption, especially that derived from animals.

Urban life-styles. Urbanisation is strongly associated with the increasing consumption of fat in developing countries. This a general phenomenon that is part of an overall change of food habits. In Bangladesh, for instance, the consumption of fat in urban areas is eight times higher than in rural areas. The same is true in Niger, where consumption in urban areas is two or three times higher than in rural ones, depending on the season. In Niger, the FER is 7 percent among people living in rural areas, while it is 14 percent among those who have recently come from rural areas to urban areas. It is 19 percent among individuals permanently living in Niamey, Niger's capital.

A survey carried out in Brazil in 1974-75 indicated a clear association between urbanisation and fat consumption independent of climatic regions. The increased FER which is commonly seen as areas become cities is striking since energy intakes tend to decrease with urbanisation, due to lower levels of physical activity.

Other factors. The physical environment, local availability of fats and oils, food habits and the level of education are other factors which affect the level of fat consumption. Sociological and individual factors also affect fat consumption. Food Balance Sheets indicate that during the last decade, a decrease in the availability of fats of animal origin, particularly in many developed countries in North Europe, North America and Oceania began. Food consumption surveys confirm this evolution

Analyses of fat consumption data show that in the most developed countries individuals belonging to the lowest socioeconomic groups consume more fatty foods. Studies show that men consume more fatty foods than women and young people eat more fatty foods than the elderly. All of these factors affect the total quantity of fat consumed. Furthermore, in developed countries food choices are increasingly linked to the type of fat contained in food.

According to FAO Food Balance Sheets, visible fats provide nearly half of the available fat in the world. Furthermore, oils of vegetable origin contribute a markedly higher percentage of visible fat than those of animal origin. The percentage of visible fats of animal origin is higher in developed countries, except North America. From data describing 165 countries, the availability of animal fats has decreased or remained at the same level in 102 countries. In the United Arab Emirates, Norway, Finland, the United Kingdom, Ireland, Canada, Australia, Denmark, the United States and the Netherlands the consumption of animal fats has been reduced. This trend became sharper at end of the 1980s.

In contrast, the sharpest increase (8 to 17 grams per caput per day) has been recorded in the eastern European countries, Cuba, Belgium, Luxembourg, Italy, France and Cape Verde. Worldwide, vegetable sources supplied 24 grams of oil per caput per day in 1990, while animals provided 6 grams of visible fat per caput per day. Out of the 165 countries, all but 12 countries have experienced increases in their vegetable oil availability since

1961. In 1990, 65 countries had more than 30 grams per caput per day of vegetable oil and 6 other countries had more than 60 grams per caput per day.

While each country has particular varieties of oil, only certain oils are important on a global level. Among the major commercial oils, the supplies of soya bean oil have increased from 2.2 to 7.0 grams per caput per day, while the availability of sunflower oil has grown from 1.3 to 3.5 grams per caput per day. Supplies of rapeseed oil increased from 0.9 to 3.4 grams per caput per day and the availability of palm oil rose from 0.9 to 2.9 grams per caput per day. Peanut oil supplies remained almost the same (2 grams per caput per day) during the last 3 decades.

Of particular interest is the content of essential fatty acids. Soybean oil contains less linoleic acid than sunflower seed oil. However, both oils provide adequate amounts of linoleic acid. On the other hand, essential fatty acids from the n-3 family may be supplied from vegetable oils containing α-linolenic acid, such as canola and soybean oils, and from fish oils containing the longer-chain fatty acids. Fish oils do not appear as separate items in food balance sheets because they are consumed as part of the fish in many countries. Only a few countries show that significant amounts of fish oils are available. Seven countries have more than 2 grams per caput per day.

PROCESSING AND REFINING EDIBLE OILS

Processing can remove the components of edible oils which may have negative effects on taste, stability, appearance or nutritional value. To the extent possible, processing should preserve tocopherols and prevent chemical changes in the triacyglycerols.

Small Scale Production

Rural oil extraction usually occurs near the areas of raw material production. This provides the smallscale processor with access to raw materials, helps to ensure that perishable oil crops are processed quickly, and reduces transport costs. For rural communities and the urban poor, unrefined vegetable oils contribute significantly to the total amount of oil consumed. Crude oils are affordable to low-income groups and serve as important sources of β-carotene and tocopherols.

To maintain the quality of the raw material, care is needed during and after the harvesting of oilbearing fruits that are perishable and susceptible to fat breakdown. Bruising of fresh palm fruits accelerates lipase activity leading to fat degradation. Oil-bearing crops such as sheanuts are prone to mould infestation during storage. This is curtailed by heat treatment: steaming or boiling, coupled with sun-drying to reduce the moisture content.

Storage. The moisture content of oil seeds and nuts influences the quality of raw materials over time. In most rural operations, sun-drying reduces the moisture content of oil seeds to below 10 percent. Adequate ventilation or aeration of the seeds or nuts during storage ensures that low moisture levels are maintained and microbial development is avoided. This is important in the storage of groundnuts which are highly susceptible to aflatoxin contamination through the growth of Aspergillusflavus.

Since aflatoxins and pesticides are not removed by rural extraction techniques, microbial contamination and the application of insecticides should be avoided. There is a need for storage practices which are affordable and available to the small-scale processor. Perishable raw materials such as palm fruits should be processed as soon as possible after harvesting. In humid developing countries, the sun-drying of oil seeds with a high moisture content, such as mature coconut, is slow and inefficient. Such conditions promote mould growth which results in high free fatty acid levels and poor organoleptic qualities. Coconut oil for human consumption should be obtained soon after harvest.

Pre-treatment. The first operation after harvesting involves sterilisation and heat treatment by steaming or boiling, this inactivates lipolytic enzymes which could cause rapid degradation of the oil and facilitates the pulping of the mesocarp for oil extraction. "Sterilised" palm fruits are pulped in a wooden pestle and mortar or mechanised digestor. Decortication or shelling separates the oil-bearing portion of the raw material and eliminates the parts that have little or no nutritional value. Small-scale mechanical shellers are available for kernels and nuts although manual cracking is still prevalent.

Most oil seeds and nuts are heat-treated by roasting to liquify the oil in the plant cells and facilitate its release during extraction. All oil seeds and nuts undergo this treatment except palm fruits for which "sterilisation" replaces this operation. To increase the surface area and maximise oil yield, the oil-bearing part of groundnuts, sunflower, sesame, coconut, palm kernel and sheanuts is reduced in size. Mechanical discattrition mills are commonly used in rural operations.

Extraction. In oil extraction, milled seed is mixed with hot water and boiled to allow the oil to float and be skimmed off. The milled oil seed is mixed with hot water to make a paste for kneading by hand or machine until the oil separates as an emulsion. In groundnut oil extraction, salt is usually added to coagulate the protein and enhance oil separation. A large rotating pestle in a fixed mortar system can be powered by motor, humans or animals to apply friction and pressure to the oil seeds to release oil at the base of the mortar. Other traditional systems used in rural oil extraction include the use of heavy stones, wedges, levers and twisted ropes.

For pressing, a plate or piston is manually forced into a perforated cylinder containing the milled or pulped oil mass by means of a worm. The oil is collected below the perforated chamber. A variety of mechanical expellers have been designed. The pre-heated raw material is fed into a horizontal cylinder by a wormshaft. By means of an adjustable choke, internal pressure which is built up in the cylinder ruptures the oil cells to release the oil.

Dehydration. By boiling in shallow pans, traces of water in crude oil are removed after settling. This is common in all rural techniques which recognise the catalytic role of water in the development of rancidity and poor organoleptic qualities.

Pressed cakes. A by-product of processing, the pressed cake, may be useful depending on the oil extraction technique applied. Cakes from water-extracted oil are usually depleted of nutrients. Other traditional techniques, for instance, those used for groundnut and copra ensure that the by-products, if handled with care, are suitable for human consumption.

Traditional technologies. In many countries, traditional processes for producing oil are very important, especially among communities which have easy access to raw oleaginous materials. Traditional processing tends to be environmentally sound and the skills required are family or group activities, involving women in particular. In a changing industrial atmosphere, these positive features have been outweighed by the negative aspects of traditional processing such as small production capacities, poor economies of scale, high expenditures of energy and time, and the cost of transporting oils to markets.

Large-scale Production

Storage. Many steps in industrial processing find their origin in the traditional processes. In large-scale operations, oilseeds are dried to less than 10 percent moisture. They may be stored for prolonged time periods under suitable conditions of aeration with precautions against insect and rodent infestation. Such storage reduces mould infection and mycotoxin contamination and minimises biological degradative processes which lead to the development of free fatty acids and colour in the oil. Oil-bearing fruits such as olive and palm are treated as quickly as possible. Palm is sterilised as a first step in processing. Adipose tissues and fish-based raw materials (that is, the body or liver) are rendered within a few hours by boiling to destroy enzymes and prevent oil deterioration.

Processing. Oilseeds are generally cleaned of foreign matter before dehulling. The kernels are ground to reduce size and cooked with steam, and the oil is extracted in a screw or hydraulic press. The pressed cake is flaked for later extraction of residual fat with solvents such as "food grade" hexane. Oil can be directly extracted with solvent from products which are low in oil content, that is, soybean, ricebran and corn germ. After sterilisation, oil-bearing fruits are pulped (digested) before mechanical pressing often in a screw press. Palm kernels are removed from pressed cakes and further processed for oil. Animal tissues are reduced in size before rendering by wet or dry processes. After autoclaving, tissues of fish are pressed and the oil/water suspension is passed through centrifuges to separate the oil.

Oil Refining. Refining produces an edible oil with characteristics that consumers desire such as bland flavour and odour, clear appearance, light colour, stability to oxidation and suitability for frying. Two main refining routes are alkaline refining and physical refining (steam stripping, distillative neutralisation) which are used for removing the free fatty acids. The classical alkaline refining method usually comprises the following steps:

— Degumming with water to remove the easily hydratable phospholipids and metals.

— Addition of a small amount of phosphoric or citric acid to convert the remaining non-hydralable phospholipids (Ca, Mg salts) into hydratable phospholipids.

— Neutralising of the free fatty acids with a slight excess of sodium hydroxide solution, followed by the washing out of soaps and hydrated phospholipids.

— Bleaching with natural or acid-activated clay minerals to adsorb colouring components and to decompose hydroperoxides.

— Deodorising to remove volatile components, mainly aldehydes and ketones, with low threshold values for detection by taste or smell. Deodorisation is essentially a steam distillation process carried out at low pressures (2-6 mbar) and elevated temperatures.

For some oils, such as sunflower oil or rice bran oil, a clear table product is obtained by a dewaxing step or crystallisation of the wax esters at low temperature, followed by filtration or centrifugation. The alkaline neutralisation process has major drawbacks, the yield is relatively low and oil losses occur due to emulsification and saponification of neutral oil. Also, a considerable amount of liquid emuent is generated. The soaps are generally split with sulphuric acid to recover free fatty acids along with sodium sulphate and some fat-containing acid water steam.

In physical refining, the fatty acids are removed by a steam distillation (stripping) process similar to deodorisation. The low volatility of fatty acids (depending upon chain length) requires higher temperatures in physical refining than those required for only deodorisation. In practice, a maximum temperature of 240-250°C is sufficient to reduce the free fatty acid content to levels of about 0.05-0.1 percent. A prerequisite for physical refining is that phosphatides be removed to a level below 5 mg phosphorus/kg oil.

In the classic refining process, this level is easily achieved during the neutralisation stage, but special degumming processes may be required for physical refining of high-phosphatide seed oils. These procedures rely on improved hydration of phospholipids by intimate contact of the oil with an aqueous solution of citric acid, phosphoric acid andlor sodium hydroxide, followed by bleaching.

It is unlikely that the mild reaction conditions during degumming and neutralisation will induce any significant undesirable changes in the oil composition. On the contrary, several impurities including oxidised components, trace metals and colouring materials are partially removed by entrainment with the phospholipids and soapstock. These impurities are further reduced during bleaching. Neutralisation also contributes significantly to the removal of contaminants such as aflatoxin and organophosphorous pesticides.

Organochlorine pesticides and polycyclic aromatic hydrocarbons, if present, must be removed during the deodorisation/stripping stage and by active carbon treatment. Some loss of tocopherols and sterols during alkaline neutralisation usually occurs, however, under well-controlled conditions (minimising air contact) this need not exceed 5-10 percent.

The possibility of negative effects of high temperatures during deodorisation and stripping has evoked concern. In several studies, extreme conditions of temperature and time (even with free access of air) were applied to produce significant quantitative results. However, the results of model studies should be related to practical processing conditions. As early as 1967-1979, the German Society for Fat Research (DGF) defined upper limits for deodorisation conditions. Good manufacturing practice also implies the following: the use of stainless steel equipment; the careful deaeration at < 100°C before heating to the final stripping temperature; the use of oxygen-free steam; and strict feedstock specifications.

Investigations in which oil was maltreated under extreme conditions determined the effects of temperature (240-300°C) and time (30-180 min.) during "physical refining" of soybean oil (degummed with phosphoric acid and lightly bleached, but still containing 20 mg P, 0.35 mg Fe and 0.05 mg Cu per kg oil). Time also has a significant effect. At 280-300°C, there was evidence for appreciable inter- or intra-esterification (increase in the content of saturated fatty acids at the 2-position of the triacylglycerols); substantial amounts of conjugated fatty acids were formed as well. The shaded areas indicate the usual range of processing conditions required for physical refining (270°C for 30 mini 250°C for 1 h; 240°C for 2 h; 220°C for 3 h). Under these conditions, all changes induced by the high temperature treatment appear to be relatively minor.

Cis-trans isomerisation. One of the most sensitive parameters used to detect chemical changes resulting from severe processing conditions is cis-trans isomerisation, especially of linolenic acid. The most complete study in this area was made by Eder, who investigated the formation of geometric isomers in various oils on the laboratory, pilot plant and production scales. On a laboratory scale, with unbleached soyabean oil at 240°C, the formation of C18:3 isomers (determined by GLC) was insignificant (less than 1 percent, even after 5 h, against 3 percent at 260°C).

Dimensahon and polymensahon. Quantitative data on the formation of polymeric compounds from bleached soybean oil are included in Figure 4. Up to about 260°C, their rate of formation appears to be low; above 260°C, the increase is more rapid. Similar trends were observed by Eder. A rapid increase of the amount of polymeric triacylglycerols was only observed at 270°C. Deodorisation of soybean oil on a commercial scale (2 x 51 min at 240°C) resulted in an increase of 0.5 percent to 0.8 percent polymeric triacylglycerols.

When the temperature was increased to 270°C in one of the deodoriser trays, 1.5 percent polymers was found. This suggests that the content of di- and polymeric triacylglycerols do not normally exceed l percent by weight in properly refined oils and fats. Strauss, Piater and Sterner carried out toxicological studies on mice fed concentrates (24 or 96 percent) of dimeric (incl. polymeric) triacylglycerols isolated from soybean oil that had been deodorised for 3.5 h at 220°C and 1 h at 270°C and contained 1.5 wt percent of dimers.

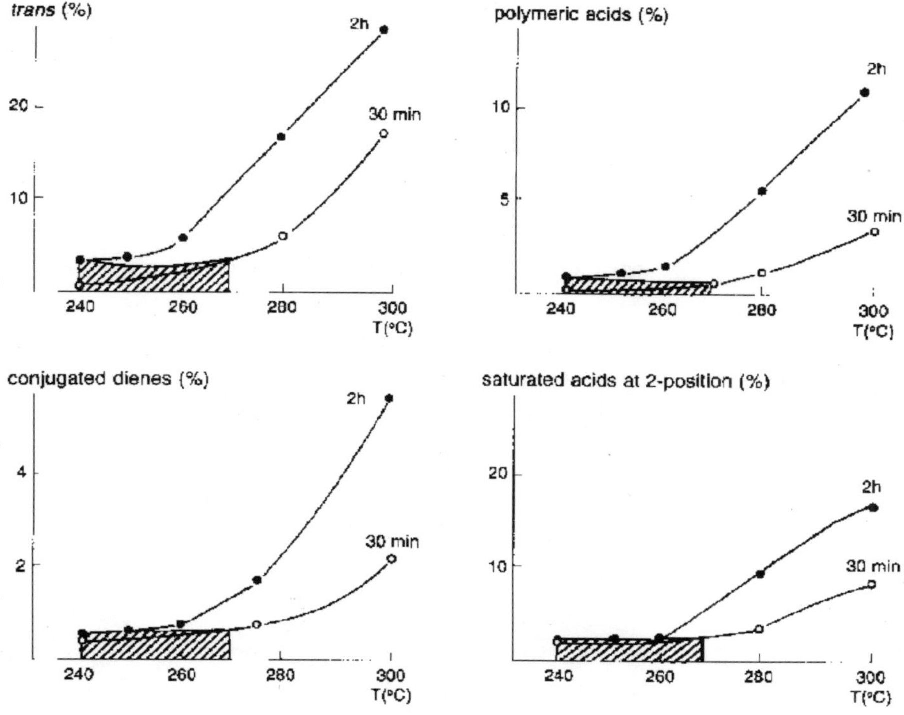

Figure 4: Formation of various artefacts in soybean oil under extreme conditions of heat treatment

The acute toxicity was found to be low with an LD 50 above 18 g/kg body weight. Long term administration of the dimer concentrate (12 months, 15 wt percent of the 24 percent concentrate in the diet) did not reveal any significant differences compared with the control group. The absorption of dimeric fatty acids was found to be poor. Therefore, the presence of small amounts of dimers and polymers in processed oils does not seem to present a physiological problem.

During deodorisation or physical refining, volatile components are removed from the oil by the combination of high temperature, low pressure and stripping action of inert gas (steam). The degree of removal depends on the physical properties of the components (especially vapour pressure) and on the temperature and volume of steam passed through the oil. Some physical losses are highly desirable, for example, the removal of off-flavours, pesticides and polycyclic aromatic hydrocarbons, if present. Other losses of nutritionally valuable components, such as tocopherols and sterols, are potentially undesirable.

Quantitative data on the composition of deodoriser distillates and the removal of various types of contaminants from oils during deodorisation were reported. During high temperature deodorisation or physical refining, especially, the concentrations of organochlorine pesticides, if present in the bleaching oil, are reduced to very low levels. Polycyclic aromatic hydrocarbons (PAM) have been a reason for concern since they were detected in some types of crude oils and fats in the 1960s. An example of this is coconut oil from copra dried with non-purified smoke gases. Depending upon the number of aromatic rings, polycyclic aromatic hydrocarbons are classified as either light (3-4 rings) or heavy (5 or more rings). Some of these compounds have proven carcinogenic properties, for example, benz-α-pyrene. Light polycyclic aromatic hydrocarbons can be removed by deodorisation or physical refining, whereas heavy polycyclic aromatic hydrocarbons can be removed only by adsorption onto activated carbon. This treatment, which can be combined with the bleaching process, is effective in reducing the amount of polycyclic aromatic hydrocarbons to acceptable levels.

Some loss of tocopherols and sterols by evaporation during high temperature deodorisation and physical refining is inevitable. However, they have higher molecular masses and lower volatilities than free fatty acids and light polycyclic aromatic hydrocarbons. Consequently, the losses of tocopherols and sterols need not be severe if the processing conditions are well-chosen. Extreme conditions applied in some model studies do induce more drastic effects. After 2 hrs at 300°C (a drastic treatment), the tocopherols and sterols almost completely disappeared, whereas the actual reduction during physical refining at 240°C for 120 min is only about 15-20 percent. The total refining losses (including pretreatment) are about 25-35 percent. Many investigators report similar data for a variety of oils. The losses tend to be higher in physical refining than in alkaline refining because of higher stripping temperatures. Under extreme conditions, some isomerisation of β-sitosterol can occur and individual tocopherols (α,β, τ, δ) and sterols may behave differently when exposed to high temperatures. However, these phenomena are of minor significance under more realistic conditions. The percentage composition of the tocopherol and sterol fractions remains essentially unchanged during processing.

For refined oils such as sunflowerseed, cottonseed and rapeseed oil, an upper limit of 3035 percent loss of tocopherols throughout the entire processing would still satisfy the generally accepted criterion of octocopherol equivalents/linoleic acid ≥ 0.6 mg/g. Soybean oil, which is high in γ-tocopherol and consequently well-protected in vitro, is relatively low in α-tocopherol and cannot reach this ratio. The, β-carotene in palm oil is another valuable component that needs consideration in the refining process. Special processes for retention are being designed. Olive oil and sesame oil are used in unrefined form since a specific taste is expected by consumers.

Fat Modification Processes

Hydrogenation. Hydrogenation of edible oils and fats has been applied on a large scale since the beginning of this century. The process is carried out in a three-phase system (hydrogen gas, liquid oil and solid catalyst) at temperatures ranging from about 120°C to about 220ºC max. in the final stages of the reaction. The catalyst consists of small nickel crystallites supported by an inorganic oxide, usually silica or alumina. After the reaction, the catalyst is filtered off and any traces of residual nickel are removed in post-refining to a level of about 0.1 mg/kg or below. Hydrogenation is a series of consecutive reactions following pseudo first order reaction kinetics:

$$18{:}3 \xrightarrow{K_3} 18{:}2 \xrightarrow{K_2} 18{:}1 \xrightarrow{K_1} 18{:}0$$

in which K_3, K_2 and K_1 are the reaction rate constants for linolenic, linoleic and oleic acid, respectively. The K_3 is much larger than K_2 and K_1. In nearly all hydrogenations, linolenic acid is transformed into compounds which are less saturated. Depending upon the reaction conditions, the so-called Selectivity Ratio (K_2/K_1) can change significantly; for example, for nickel catalysts it varies from approximately 10 at a low temperature to 50 or 100 at high temperature. A high Selectivity Ratio implies that relatively little saturated acid is formed and that monounsaturated fatty acids are the dominant reaction product.

Apart from reduction of unsaturation, isomerisation of double bonds also takes place during hydrogenation: geometric isomerisation (cis-trans) and positional isomerisation. The mechanisms of hydrogenation and isomerisation are strongly interrelated. Initially, a half-hydrogenated intermediate is formed in which the molecule adsorbed to the catalyst surface by a single bond can freely rotate. Addition of a second hydrogen atom results in saturation, whereas abstraction of a hydrogen atom from the half-hydrogenated intermediate to the nickel surface yields either the original molecule or a positional or geometric isomer.

Hydrogenation of polyenoic fatty acids occurs, at least partially, through conjugated isomers which are highly reactive and, therefore, rapidly converted into cis or trans monoenoic acids without accumulating. Of the total amount of trans fatty acids present in hydrogenated oils and fats, the greater part by far are trans monoenes. Due to the important role hydrogenation plays in the production of plastic fats, trans fatty acids may be found in considerable amounts in many products. The amount of cis, trans and trans, cis dienes is much smaller, whilst the level of trans, trans dienes seldom exceeds 1 percent.

Apart from hydrogenation, there are two other important fat modification technologies. The first is interesterification, the random rearrangement of the fatty acids in the triglyceride molecule under the influence of a moderately alkaline catalyst. This modifies the melting behaviour of the fat without changing the nature of the fatty acids. The second is fractionation, the controlled separation of oil/fat fractions by low temperature (dry fractionation), or solvents (solvent fractionation). In this process, there are no changes in the chemical nature of the fatty acids.

Palm oil is fractionated into palmolein and palmstearin. Interrelationships of fat modification technologies. Industry uses various oils and fats interchangeably while retaining constant quality. Generally, the least expensive combination of raw materials compatible with the required quality is chosen. Hydrogenation greatly extends the number of fats available with specified melting behaviour and this increases interchangeability and lowers costs. In non-hydrogenation situations, a combination of interesterification, fractionation and selection of the starting oil may be acceptable solutions for limiting the formation of isomers in terms of product quality, however, the costs are higher. While the actual specific modification, for example, hydrogenation or interesterification, is relatively inexpensive, the costs of losing flexibility can be significantly higher. The processes for oil modification may change as new compositions become available through plant biotechnology.

Other considerations

Storage, transport and packaging of oils. Oils and fats must be protected against oxidative deterioration, contamination with water, dirt, or other fats, absorption of foreign odours and tastes, thermal deterioration, and entry of foreign substances from packaging and lining materials. Temperature, oxygen pressure, oxidation products, trace metals, oxidative and lipolytic enzymes, reduction in natural antioxidants and visible and ultraviolet light are all factors in such deterioration. The use of low storage temperatures, nitrogen or vacuum packaging; the avoidance of copper, copper alloys and iron as construction material of storage vessels; and the use of synthetic or natural antioxidants and metal sequestrants as additives, work to prevent deterioration of oil during storage.

Selection of oil processing technology. The use of oil processing technology and its application in products is influenced by several factors. For example, the demand for triacylglycerols with specific fatty acids in the 1, 2 and 3 positions of the molecule can be met through enzymatic interesterification which uses lipases as catalysts in the interesterification process. The more readily absorbed fatty acid in the 2-position can provide for specific triacylglycerols with medical benefits. For example, essential fatty acids can be supplied to patients with various types of fat malabsorption or energy can be given to infants with palmitic acid in the 2-position. Another is the increased use of physical refining because of consumer pressure for less "chemical" processing.

Selected Uses of Fats and Oils in Food

Fats are the primary constituents of margarines, butterfat, shortenings, and oils for salad and cooking. In addition to the visible fat contained in food, fats and oils are found in high quantities in many bakery goods, infant formulas, and dairy products and some sweets. Oils, butter or margarine are sometimes used directly on food.

Cooking Oils

The major use of cooking oil is in frying, where it functions as a heat transfer medium and contributes flavour and texture to foods. One requirement of a cooking oil is that it be stable under the very abusive conditions of deep-fat frying, namely, high temperatures and moisture. In general, oil should be kept at a maximum temperature of 180°C during frying. Frying food at a temperature which is too low results in increased fat uptake. Water, which is contributed by the foods that are fried in an oil enhances the breakdown of fatty acids which occurs during heating. Hydrolysis results in a poor-quality oil that has a reduced smokepoint, darkened colour and altered flavour. During heating, oils also polymerize, creating a viscous oil that is readily absorbed by foods and that produces a greasy product. The more saturated (solid) the oil, the more stable it is to oxidative and hydrolytic breakdown, and the less likely it is to polymerize.

Oils rich in linolenic acid, such as soybean and canola oils, are particularly susceptible to these undesirable changes. When soybean oil is partially hydrogenated to reduce the linolenic acid from about 8 percent to below 3 percent, it is a relatively stable frying oil and it is used in processed fried foods, pan and griddle frying, and sauces. Stability is increased by using cottonseed, corn oil, palm oil or palmolein or by more hydrogenation of soybean oil.

Foods that are fried and stored before eating, for example, snack products, require an even more stable oil. More saturated oils improve stability, however, if the frying fat is solid at room temperature it will produce a dry dull surface that is undesirable on some fried products. When oils are used continuously, as in restaurants, a frying fat which can withstand very heavy use is needed. In these cases, more solid shortenings are used to maximize the stability of the fat for many hours of frying.

Frying oils made from sunflower and safflower have lower stability because of their high polyunsaturated fatty acids and low g -tocopherol content; however, high-oleic safflower and sunflower oils that have been genetically developed are suitable frying oils.

For optimal use of cooking oils, it is necessary to distinguish between different frying conditions. The most important parameters to be monitored are duration of use and nature of the foods to be fried. If food fats enter the frying oil, food components could destabilize the oil and the water content of the material could influence the frying operation. Whether the use is continuous or intermittent is relevant, continuous use provides a protective water vapour blanket that protects against oxidation. Finally, temperature must be considered.

Industrial use of fats and oils is usually well-monitored. The continuous operation (implying the constant addition of fresh oil) and quality requirements for the products normally ensure good quality control of the oil. In homes, where oils are normally used for much shorter periods of time and are discarded after being used once or twice, stability problems play a lesser role. Stability of frying oils is a more important factor in catering operations, where heating is intermittent and oils may be used for long periods.

Margarines

Margarines must have some crystalline structure to maintain a semisolid consistency at refrigerator and room temperatures. Sharp melting at body temperature is needed so that the margarine will melt rapidly in the mouth leaving no waxy feeling.

Oleic acid melts at 16°C, whereas elaidic acid melts at 44°C, so that the presence of some trans isomers may vastly raise the melting point and stability of a product. Stick-type margarines contain 1029 percent trans fatty acids while tub-type margarines have 10-21 percent trans fatty acids. In addition to partial hydrogenation, the correct consistency of a margarine can be obtained by blending soft and hard fats. Lower fat spreads, for example, 40 percent or 60 percent fat, contain less trans fatty acids.

Another important feature in solidifying oil for margarines is the type of crystal formed. Fats are polymorphic, that is, they are capable of forming several different types of crystals. The a crystals are the tiniest, forming a smooth but unstable crystal. The b ‘ crystals are medium in size, still they are generally desirable in margarines because they impart a smooth texture, are fairly stable, and ensure plasticity of the product. The largest crystals are the b type which are stable and grainy, and generally undesirable. Also, the , b form is readily converted to a hard and brittle structure. Products such as liquid shortenings and coating fats sometimes require the ,b crystal.

The lengths of fatty acids and their positions on the glycerol backbone determine the type of crystal formed. The triacylglycerols in a certain fat or solidified oil always form the same type of crystals unless other ingredients are added to alter crystal formation. To produce a margarine with enhanced 'b' stability, it is necessary to have a variety of triacylglycerols with different fatty acid chain lengths. Palm and hydrogenated cottonseed oils contain a fair amount of C16:0 and can be added to other oils to enhance b ' structure.

Shortenings

Shortenings are semisolid fats that impart a "short" or tender quality to baked goods, enhance the aeration of leavened products, and promote a desirable grain and flavour. They coat the gluten proteins of flour which prevents toughness. In contrast, toughness is desirable in yeast-raised products to give a chewy texture. For products with characteristics between breads and cakes such as doughnuts, shortening modifies the gluten and adds richness to the product. In baked goods, shortenings are used specifically to leaven, cream and lubricate. In icings and fillings, the fats help to form tiny air bubbles that create a light and fluffy structure. Shortenings used as stable frying fats provide a heating medium, and their crystalline structures are not important.

The requirements of fats with properties for shortening are fairly specific depending upon the food in which they are used. Bakery shortenings should have as wide a plastic range as possible, that is, the melting behaviour should remain constant over a specified temperature range, often 24-42°C. This quality allows the fat to be easily manipulated without melting at room temperature and enhances its mixing ability. A wide plastic range is accomplished by blending a partially hydrogenated stock with a fully hydrogenated oil such as soybean (b-crystal) or cottonseed and palm (b-crystal). The b-crystal is often preferred because it results in a creamier texture.

Salad Oils

A major use of salad oils is in pourable salad dressings. Traditional salad dressings, some of which are emulsified, consist of a two-phase system of oil and water with 55-65 percent oil. A salad oil coats the salad ingredients, spreading the flavour of the dressing that improves the palatability of the salad. The other major use of salad oils is in mayonnaise and thick salad dressings, which contain 80 and 35-50 percent oil, respectively. The oil in mayonnaise is responsible for viscosity, whereas the oils in thick salad dressings help to modify the mouthfeel of the starch paste that thickens the product.

A salad oil must not contain solid crystals that, when refrigerated, would impart a waxy, tallowy texture, would break the emulsion formed between water and oil, or would give the product a cloudy appearance. Oils can be winterized, a process that removes solid crystals formed at refrigerator temperatures.

Typically, unhydrogenated or partially hydrogenated soybean, canola, winterized cottonseed, safflower, sunflower, and corn oils are used. Olive oil has a unique flavour, although it forms crystals at refrigerator temperature, it is often served at room temperature as a salad oil.

Medium-chain triglycerides (MCT)

In addition to the common dietary fats, lipid fractions such as medium-chain triglycerides (MCT oil) are used in specialized therapeutic preparations. MCT oil is a fraction of coconut oil containing fatty acids of 8-10 carbon atoms in triacylglycerols. MCT oil is used in formulas for enteral feeding and in diets for patients with malabsorption syndromes.

References

Carroll, K.K. and Khor, H.T. (1975). Dietary fat in relation to tumorigenesis. *Progress in Biochemical Pharmacology*. 10: 308-353.

Miller, W.C., Lindeman, A.K., Wallace, J. and Niederpruem, M. (1990). Diet composition, energy intake, and exercise in relation to body fat in men and women. *American Journal of Clinical Nutrition* 52:426-430.

Reed, D., McGee, D., Yano, K. and Hankin, J. (1985). Diet, blood pressure and multicollinearity. *Hypertension*. 7: 405-410.

Rolls, B.J. and Shide, DJ. (1992). The influence of dietary fat on food intake and body weight. *Nutrition Reviews*. 50: 283290.

7

Causes of Malnutrition

> We, the Ministers and Plenipotentiaries representing 159 nations... declare our determination to eliminate hunger and to reduce all forms of malnutrition. Hunger and malnutrition are unacceptable in a world that has both the knowledge and the resources to end this human catastrophe.

These are the opening sentences of the World Declaration on Nutrition produced by the FAO and World Health Organization (WHO) International Conference on Nutrition (ICN) held in Rome in December 1992. That important conference reviewed the current nutrition situation in the world and set the stage for markedly reducing these unacceptable conditions of humankind. Reaching the ICN goal is possible. Most of the work will need to be done in the developing countries by their own people. However, cooperative work across nations and across disciplines is also essential.

The ICN declaration goes on to state:

1. ... We recognize that globally there is enough food for all and that inequitable access is the main problem. Bearing in mind the right to an adequate standard of living, including food, contained in the Universal Declaration of Human Rights, we pledge to act in solidarity to ensure that freedom from hunger becomes a reality. We also declare our firm commitment to work together to ensure sustained nutritional well-being for al] people in a peaceful, just and environmentally safe world.
2. Despite appreciable worldwide improvements in life expectancy, adult literacy and nutritional status, we all view with the deepest concern

the unacceptable fact that about 780 million people in developing countries - 20 percent of their combined population - still do not have access to enough food to meet their basic daily needs for nutritional well-being.

3. We are especially distressed by the high prevalence and increasing numbers of malnourished children under five years of age in parts of Africa, Asia and Latin America and the Caribbean. Moreover, more than 2 000 million people, mostly women and children, are deficient in one or more micronutrients; babies continue to be born mentally retarded as a result of iodine deficiency; children go blind and die of vitamin A deficiency; and enormous numbers of women and children are adversely affected by iron deficiency. Hundreds of millions of people also suffer from communicable and non-communicable diseases caused by contaminated food and water. At the same time, chronic non-communicable diseases related to excessive or unbalanced dietary intakes often lead to premature deaths in both developed and developing countries.

The Scale of the Problem

Protein-energy malnutrition (PEM), vitamin A deficiency, iodine deficiency disorders (IDD) and nutritional anaemias - mainly resulting from iron deficiency or iron losses - are the most common serious nutritional problems in almost all countries of Asia, Africa, Latin America and the Near East.

Nutrition and development: a global assessment, prepared by FAO and WHO for the ICN, reviewed all available current information on the prevalence of hunger and malnutrition and provided a global estimate for the various regions of the world. FAO updated the estimates of the chronically undernourished population of the world for the Sixth World Food Survey and in preparation for the World Food Summit, and WHO updated the estimates for iodine, vitamin A and iron deficiencies in 1995. The figures suggest that one of every five persons in the developing world is chronically undernourished, 192 million children suffer from PEM and over 2 000 million experience micronutrient deficiencies. In addition, diet-related non-communicable diseases such as obesity, cardiovascular disease, stroke, diabetes and some forms of cancer exist or are emerging as public health problems in many developing countries.

While these numbers and trends are alarming, progress has been made in reducing the prevalence of nutritional problems, and many countries have been remarkably successful in addressing the issues of hunger and malnutrition. For the developing countries as a whole there has been a consistent decline since the early 1970s in the proportion and absolute number of chronically undernourished people. From 1969 to 1971 approximately 893 million people were chronically undernourished, compared with 809 million from 1990 to 1992; these figures represent a drop from 35 to 20 percent of the population of these countries. The current - and achievable - challenge is to build upon and accelerate the progress that has been made.

FAO and WHO data indicate improvements of the nutritional situation in Asia and Latin America from 1980 to 1990 but a deterioration in sub-Saharan Africa. Although the prevalence of underweight children remained virtually unchanged in sub-Saharan Africa during that decade (increasing from 29 to 30 percent), the prevalence rates are much better than in South Asia, where about 59 percent of children - almost twice the prevalence in Africa - were underweight in 1990. In the same year, in total numbers, five times as many children were underweight in South Asia (101 million) as in sub-Saharan Africa (19.9 million).

Many nutritional statistics show the numbers of persons who have overt evidence of a deficiency. However, "at risk" populations are not often identified. In nutrition, as in public health, people considered at risk of developing malnutrition should be among the primary concerns. Prevention becomes more feasible and cost effective if groups at risk are identified and the causes of malnutrition are clearly understood.

One of the most dramatic aspects of the global nutrition situation is the extent of famine, hunger and starvation. While good progress has been made in averting famine, especially in Asia, these horrifying conditions persist throughout the world. Their occurrence is commonly attributed to drought and other natural disasters, but war, civil unrest and political instability have far greater importance. In the mid-1990s, hunger and malnutrition resulting from civil strife are serious problems in many parts of the world including Europe (particularly former Yugoslavia), Asia (for example, Afghanistan), the Near East (Iraq) and most extensively Africa. Tragically, civil strife often affects not only the countries in turmoil but also those that provide hospitality to the refugees who flee their homes in terror.

In mid-1994, the United Republic of Tanzania accepted about 500 000 refugees from Rwanda, most of them in less than one week. Their arrival more than doubled the population of the resource-poor region, which welcomed them as best it could. The influx placed overwhelming pressure on local resources and necessitated a major international effort to prevent an increase in nutrition and health problems among the local people as well as to contain these problems among the refugees.

Nutrition Improvement: Nature and Evolution

Data from around the world show that the causes underlying most nutrition problems have not changed very much over the past 50 years. Poverty, ignorance and disease, coupled with inadequate food supplies, unhealthy environments, social stress and discrimination, still persist unchanged as a web of interacting factors which combine to create conditions in which malnutrition flourishes. However, what does change greatly is the approach to tackling malnutrition. Each decade or so witnesses a new dominant framework, paradigm, panacea or quick fix claimed to be capable of reducing the malnutrition problem greatly before ten years have passed.

During the 1950s and 1960s, kwashiorkor and protein deficiencies were seen as the major problems. Quick fixes such as fish protein concentrate, single-cell protein or amino acid fortification and increased production of protein-rich foods of animal origin were the strategies proposed for the control of malnutrition in the tropics and subtropics.

During the late 1960s and 1970s, the term "protein-energy malnutrition" entered the literature. Increasing protein and energy intake by children was the solution, and nutrition rehabilitation centres and applied nutrition programmes (ANPs) were offered up as sure strategies.

The 1974 World Food Conference began a decade of macroanalysis which placed first nutrition planning and then nutritional surveillance among the dominant strategies for the countries most affected. Economists began to take over from nutritionists and paediatricians as the architects of new policies, with much talk about national food security and agencies such as the World Bank stressing income generation.

In 1985 the International Monetary Fund (IMF) began to push structural adjustment, and WHO and UNICEF reinvented ANPs, which they renamed Joint Nutrition Support Programmes (JNSPs). In the early 1990s the subject of micronutrients pushed PEM to the background, as nutritionists, international agencies and universities attempted quick fixes to control vitamin A deficiency, anaemia and IDD. The micronutrient wave has not yet crested, and very large sums of money are likely to be provided by the World Bank, the United States Agency for International Development (USAID) and others to address this hidden hunger". This effort is, in part, a response to the goals set by the 1989 World Summit on Children and the 1992 International Conference on Nutrition, which include the virtual elimination of vitamin A deficiency and IDD before the turn of the century.

Increased funding is needed if improvements in nutrition are to be achieved. However, there is a danger that the limited resources available may be diverted towards the development of new quick-fix strategies for micronutrient deficiencies. Little, then, will remain for addressing the underlying and basic causes of malnutrition. The quick fix addresses only the immediate causes of a problem, scratching the surface and providing no sustainability.

It is well recognized that inappropriate development strategies also contribute to the underlying causes of hunger in many countries. Policy reform and the institution of appropriate development and macroeconomic policies are advocated by many economists to improve nutrition. The ICN also emphasized that developing countries must work to ensure that development policies and projects are designed to include nutrition improvement objectives. Furthermore, in the low-income food-deficit countries, where most of the world's malnourished people live, economic growth and poverty alleviation must be based on better development of agricultural resources and improvement of food supplies. This approach should promote sustainable development, expand employment opportunities and improve access to food by the poor. Free and fair trade is clearly important for stimulating economic growth, and the prices for primary and processed agricultural products must be adequate to ensure sustained development. The primary producers must receive fair prices for their products, labour and use of resources.

It has to be recognized that inappropriate application and transfer of technology and even aspects of certain development projects can have negative as well as positive consequences for health and nutrition in poor countries. It is important that such possible negative consequences be identified early and that measures be taken to offset and prevent them. It may be more important to enhance during project preparation those aspects that will have a positive impact for maximum nutritional benefits.

There is also a greater realization that the poor should be more involved in solving their own problems and that the causes of malnutrition and the different levels of society implicated vary from place to place. People should be able to ask appropriate questions relevant to their situation, at the national, local or even family level, and they should be aware of the multisectoral nature of the problem of malnutrition. They can then, together with persons from different disciplines, suggest actions that might be taken at different levels. During the past ten years a good deal has been written about local participation in development decisions and programmes. The innate wisdom of peasants, with regard to agriculture as well as other development-related matters such as health and nutritional status, has finally been widely acknowledged.

It has also been recognized that international and national policies and actions can influence nutritional status in the rural villages and city slums of developing countries. The State may determine taxes, control prices, run national institutions and oversee a legal system. Almost all of these factors influence, and some of them are influenced by, the formal and informal institutions in society. Clearly these institutions influence the causes of malnutrition. Thus the presence or absence, the relevance and the quality of formal local institutions such as agriculture advisory services, health centres, primary schools and community centres have a very important role in areas related to nutrition. But the more informal institutions can also have a role in influencing food, health and care. The most important of these is the family; others include groups of friends and religious, sporting or social groups.

The realization that malnutrition is not just a food problem has been appreciated for many years, but the concept of the importance of giving consideration to food, health, education and care is of more recent origin. It is vital that this thinking continue to develop and to move forward steadily, in the place of erratic leaps in pursuit of fashion or funding. For a healthy approach, in the next ten years, the achievements should be reassessed; old strategies that have sound logic and a successful record should be protected and supported, and new policies promoted only when needed. This approach is possible with both discipline and flexibility, and examples of its success are visible today.

A Framework for Causes of Malnutrition

Malnutrition or undesirable physical or disease conditions related to nutrition can be caused by eating too little, too much or an unbalanced diet that does not contain all nutrients necessary for good nutritional status. In this book the term malnutrition is restricted to undernutrition, or lack of adequate energy, protein and micronutrients to meet basic requirements for body maintenance, growth and development.

An essential prerequisite to the prevention of malnutrition in a community is the availability of enough food to provide for the nutrient needs of all people. For adequate food to be available, certainly there must be adequate food production or sufficient funds at the national, local or family level to purchase enough food. Availability of food, however, is just part of the picture. It is now recognized that malnutrition is only the overt sign, or symptoms, of much deeper problems in society.

Inadequate dietary intake and disease, particularly infections, are immediate causes of malnutrition. It is obvious that each person must eat an adequate amount of good-quality and safe food throughout the year to meet all nutritional needs for body maintenance, work and recreation, and for growth and development in children. Similarly, one must be able to digest, absorb and utilize the food and nutrients effectively. Poor diets and disease are often the result of insufficient household food security, inappropriate care and feeding practices and inadequate health care. It is now understood that good nutrition depends on adequate levels of all three of these factors.

Other factors can also contribute to unavailability or inadequacy of resources for afflicted families. Every rural community or society has certain natural or human resources as well as a certain potential for production. A host of factors influence what and how much food will be produced and how and by whom it will be consumed.

The proper use of resources may be affected by economic, social, political, technical, ecological, cultural and other constraints. It may be affected by lack of tools or training to use them and by limited knowledge, skills and general ability to use the resources. The cultural context is of special importance for its influence, especially at the local level, on the use of resources and the establishment and maintenance of institutions.

Malnutrition may manifest itself as a health problem, and health professionals can provide some answers, but they alone cannot solve the problem of malnutrition. Agriculturists, and often agricultural professionals, are required to ensure that enough foods, and the right kinds of food, are produced. Educators, both formal and non-formal, are required to assist people, particularly women, in achieving and ensuring good nutrition. Tackling malnutrition often requires the contributions of professionals in economics, social development, politics, government, the labour movement and many other spheres.

Promotion and Protection of Nutritional Well-being

The International Conference on Nutrition developed nine common areas for action to promote and protect the nutritional welfare of the population:

— improving household food security,

— protecting consumers through improved food quality and safety,

— preventing specific micronutrient deficiencies,

— promoting breastfeeding,

— promoting appropriate diet and healthy lifestyles,

— preventing and managing infectious diseases,

— caring for the economically deprived and nutritionally vulnerable,

— assessing, analysing and monitoring the nutrition situation.

— incorporating nutrition objectives into development policies and programmes

Addressing issues under these themes facilitates the development of a common understanding of nutrition problems by various sectors and allows a more focused approach for working towards solutions. Taking this thematic approach to nutrition problems should ensure that each of the many facets of a problem are noted, which should allow each sector or agency to assess how it can best work for improvements. These issues are discussed in detail in Part V.

The Six Ps

By shedding the sectoral perspective and adopting a multisectoral, multidisciplinary one, it is possible to see the causes of malnutrition in a different guise and to focus the development of solutions less narrowly than in the past. Each case will be different, of course, and the extent to which one cause or one area of expertise predominates will vary with the circumstances. However, six determinants of malnutrition are especially important, although none is usually the only cause of malnutrition or the only discipline that needs to be involved in nutrition strategies.

These six determinants - the six Ps - are:

— *production,* mainly agricultural and food production;
— *preservation* of food from wastage and loss, which includes the addition of economic value to food through processing;
— *population,* which refers both to child spacing in a family and also to population density in a local area or a country;
— *poverty,* which suggests economic causes of malnutrition;
— *politics,* as political ideology, political choices and political actions influence nutrition;
— *pathology* which is the medical term for disease, since disease, especially infection, adversely influences nutritional status.

Production

The production of food comes mainly from agriculture. Most countries have a ministry of agriculture and different kinds of agricultural staff whose contributions are very important to nutrition, but adequate national agricultural and food production does not guarantee good nutritional status for all people. There have been remarkable developments in agriculture in the past four decades. High-yielding varieties of the important cereals (rice, wheat and maize) have been successfully developed, and much progress has been made in increasing food yields per hectare of land. Some countries that are self-sufficient in their production of staple foods, however, still have the highest prevalence of malnutrition. Agriculturists and agriculture ministries have an absolutely vital role in improving nutritional status, but they cannot win the battle against malnutrition without action from other ministries and without other expertise. Other areas such as food safety, food losses and food storage influence the availability of food. Consideration has to be given to food demand as well as food production.

Preservation

Despite the remarkable progress made in increasing food production at the global level, approximately half of the people of developing countries do not have access to an adequate food supply. A substantial part of the food produced is lost, for various reasons, before it can be consumed. It has been estimated that about 25 percent of the grains produced are lost because of bad post-harvest handling, spoilage and pest infestation. Losses of easily perishable fruits, vegetables and roots have been estimated to be about 50 percent of what is grown. After food reaches the home, about 10 percent is lost in the kitchen. Therefore, ensuring that appropriate measures are taken to prevent food losses during harvesting, transportation, storage, processing and preservation should be an integral component of any programme for the prevention of malnutrition and the improvement of the population's access to food in developing countries. Processing can also add nutritional and economic value to foods. Adequate measures for the provision of safe and quality food should also be taken.

Population

The food available per person in a family, a district or a nation depends on the amount of food produced or purchased divided by the number of people who have access to that food. A family of eight that produces and purchases the same amount of food as a family of four has less food available per person. However, it also needs to be recognized that among producing families, larger family size can also lead to greater family productivity.

In some countries the population problem is considered to be of great importance, and overpopulation, family size and child spacing are considered important determinants of malnutrition. Demographers study population, and many countries have a government body, often in the ministry of health, responsible for family planning. Birth spacing may deserve a very high priority. However, as with production, it is naive to believe that in any country population control or successful family planning will by itself solve the problems of hunger and malnutrition.

Poverty

Poverty is often stated to be the very root cause of malnutrition. Certainly in most countries it is mainly, and sometimes only, the poor whose children suffer from severe or moderate PEM or show evidence of vitamin A deficiency. In contrast, nutritional anaemias and IDD may not be confined to the poor.

Economists are the professionals who study poverty and income and suggest economic solutions for problems of poverty which may be related to malnutrition. Most governments have a group of economists working in the ministry of finance and sometimes also in a ministry of economic planning.

The experience of many developing countries shows that a major reduction in poverty would have a significant impact on rates of PEM in most countries and communities. Efforts to reduce poverty, raise incomes, lower food prices and redistribute wealth, as well as a host of other economic policies, can have a major impact on nutrition. But just as agriculturists and demographers alone cannot solve the nutritional problems of a nation, so also economic actions alone do not usually rid a country or area of malnutrition. In some cases raised incomes have not resulted in major reductions in malnutrition and certainly have not led to its eradication.

Poverty takes many forms and is expressed in many ways. Inadequate household income is one manifestation, but poor communities and nations lack the wealth needed to build and support schools and training programmes to improve water supplies and sanitation and to provide needed health and social services.

Politics

All countries have a mechanism to create and implement policies in spheres of development. The systems differ from one country to another, but agriculture, health, education, economic and other related policies strongly influence the well-being of the people, including their nutritional status. Some governments take their obligations seriously. If government leaders take the right to freedom from want seriously, then they also respect the right to freedom from hunger, freedom from lack of health services, freedom from poor housing and so on. These conditions, however, also depend on the resources of the country. The way in which political ideology can have a significant influence on malnutrition is probably through government acting to ensure some level of equity. Equity does not imply equality, it simply means a reasonable or relatively fair access of all people to the essential resources such as housing, education, food and health care. Policies directed towards improving access of women to resources for income generation, education and health care would particularly improve the nutritional welfare of the family and children.

Pathology

This sixth P connotes disease. Physiology refers to the normal functioning of the body and its organs and cells. Pathology refers to abnormal function and to disease. Much malnutrition in the world is caused or influenced not only by shortage of food, but by disease.

The relationship between malnutrition and infection has been extensively studied and documented. There is no doubt that common infections such as diarrhoea, respiratory disease, intestinal worms, measles and acquired immunodeficiency syndrome (AIDS) are important causes of malnutrition. Certain non-infectious diseases may also be causes of malnutrition. Examples of these include a variety of malabsorption syndromes (conditions where the body does not absorb nutrients properly), many cancers and malignancies and some psychological illnesses.

Ministries of health and a variety of health personnel in the public and private sectors are responsible both for treatment of disease and for public health or preventive measures. In many countries the responsibility for government nutrition policies rests with the ministry of health, and often national institutes of nutrition fall under this ministry. Certainly health measures to prevent disease, especially infections, and also actions to provide medical care and appropriate treatment will help very much to reduce the extent of malnutrition in a country or a community. Health measures alone, however, have never been able to eliminate malnutrition totally.

A Multidisciplinary Perspective

This discussion of the six Ps, namely production, processing, population, poverty, politics and pathology, is designed to illustrate the complexity of both the underlying causes of malnutrition and the solutions. It illustrates that agriculturists, industrialists, demographers, economists, politicians and health personnel all have important roles in controlling malnutrition. It is also clear that no one ministry or single group of professionals is likely to eliminate hunger and malnutrition in society. Nutritionists, food scientists and others work across all these lines, and in a properly functioning national food and nutrition strategy they will collaborate with professionals in several of these disciplines as well as others. Achieving good nutrition may also require experts in anthropology, sociology and community development; it requires a good transport and marketing system; it benefits greatly from an education system that provides school for all, especially females, and guarantees the highest levels of literacy; and it may involve many other actors. Nutrition strategies are truly multisectoral, which may sometimes present more difficulties at the national level than at the local or community level. Community participation, with the assistance of actors from different sectors including at least agriculture, health, community development and education, will often be needed to meet the challenge of good nutrition for all.

Food Production and Food Security

A national food policy should be a part of an overall nutrition strategy with household food security for all people as a central objective. Achieving food security includes ensuring:

— a nutritionally adequate and safe food supply at both the national and household levels;

— a reasonable degree of stability in the supply of food during the year and in all years;

— access by each household to sufficient food to meet the needs of all

For all households to be food secure, each must have physical and economic access to adequate food. Each household must always have the ability, the knowledge and the resources to produce or procure the foods that it needs. Nutritionists stress also the need for the food to provide for all the nutritional requirements of the household members, which means a balanced diet providing all necessary energy, protein and micronutrients.

Beyond household food security is the need to encourage food distribution that ensures good nutritional status for all the members of the household. The right to an adequate standard of living, including food, is recognized in the Universal Declaration of Human Rights. National development policies should include food security as an objective, and achieving food security for all is an indication of success.

In nutrition there exists the paradox that while undernutrition leads to a serious set of health problems, overconsumption of food and of certain dietary components carries other risks to health. This book is particularly concerned with undernutrition.

National Food Security

Food security is often defined as access by all people at all times to sufficient food required for a healthy and active life. It is now widely accepted that most of the undernutrition in developing countries is due to inadequate intake of both protein and energy and that it is often associated with infectious diseases.

In the past, protein deficiency was overemphasized as an important nutrition problem in the world. Commercial production of relatively expensive protein-rich foods, amino-acid fortification of cereal grains, production of single-cell protein and other ventures were offered as panaceas for the world's nutrition problems. These ventures only reduced the problem of protein-energy malnutrition (PEM) by a very small degree. Thus, in the context of combating malnutrition, attempts at making small changes in the amino-acid pattern of cereal grains by means of genetic manipulation are much less useful than increasing the yields per hectare of cereals and other food crops or enabling people to purchase the foods they need.

Satisfying the energy needs of a population, which should be the first goal of a food policy, has been a relatively neglected matter. In most populations where the staple food is a cereal such as rice, wheat, maize or millet, serious protein deficiencies seldom occur except where there is also an energy or overall food deficiency. The reason is that most cereals contain 8 to 12 percent protein and are often consumed with moderate quantities of legumes and vegetables. Protein deficiencies in people consuming these diets are mainly confined to very young children suffering increased nitrogen losses because of frequent infections. However, among populations whose staple food is plantain, cassava or some other food with a low protein content, protein intakes may be a problem for greater sections of the population.

A modest increase in cereal, legume, oil and vegetable consumption by children will greatly reduce the prevalence of PEM and growth deficits for children in developing countries, especially if combined with control of infectious diseases. Breastfeeding during the first few months of life can ensure an adequate diet, whereas bottle-feeding is a major cause of diarrhoea and nutritional marasmus.

Food Availability

To nourish a population adequately, there must be a sufficient quantity and variety of good-quality and safe food in the country. Therefore, in most low-income food-deficit countries a fundamental strategy of food policy is to improve and increase food production - a domain for agriculture experts. Clearly, decision-makers in the agriculture sector need to be aware of the nutritional needs of the population and to understand the nutritional implications of their actions.

Most food in the world comes from cereals. The second largest amount of food comes from root crops, followed by legumes or pulses. In round figures, the world produces about 2 000 million tonnes of cereals, 600 million tonnes of root crops and 60 million tonnes of pulses per year. In addition, about 85 million tonnes of fats and oils and 180 million tonnes of sugar are produced worldwide each year. Developing countries produce more of all these items than do industrialized countries. In contrast, industrialized countries produce more foods of animal origin meat, milk and eggs, for example - than do the developing countries.

In the last few decades, truly remarkable advances have influenced food production. Agricultural research has developed and made available new varieties of the main cereals: rice, maize and wheat. These new varieties produce much higher yields per hectare than the old varieties. Some have a shorter period between planting and harvesting, and some are relatively resistant to disease. However, most of these new varieties require increased fertilizer use. In addition, many of the improved rice varieties and some of the wheat and maize varieties require irrigation or more water. Both of these options may be economically unfeasible for most poor farmers. In general, cultivation of improved varieties is more suitable for large, economically comfortable farms with access to agricultural inputs. It should be a major agricultural policy objective to see that more resource-poor farmers have adequate access to such inputs.

The development of these new varieties - the green revolution - has allowed much higher yields of cereals for a given area of land. As population pressure increased on arable land, the green revolution offered an alternative to the old method of increasing production, namely expanding the area of land cultivated.

Average world food production has kept pace with or very slightly exceeded the increase in world population. In round figures, 2 700 kcal are available per person per day in the world. However, the figures vary among regions; the mean for industrialized countries is around 3 400 kcal, and that for developing countries is around 2 500 kcal. Of course, average availability figures for a country mask very large differences among groups of the population.

To improve nutrition, agricultural planners should aim to expand the production of currently grown staple cereals and legumes and should promote consumption of fruits, vegetables, oilseeds and livestock products or those of small animal husbandry. Where land pressures are a constraint, particular attention should be given to maintaining a proper balance between crops and livestock.

Some countries that were major food importers in the 1960s, such as India, are virtually self-sufficient in cereal production (mainly rice and wheat). Yet in India undernutrition and malnutrition remain highly prevalent. Other countries, such as Indonesia, have become self-sufficient in rice production and have significantly reduced the prevalence of malnutrition. Some countries are far from being self-sufficient in food production yet have far less malnutrition than countries like India. For example, many Caribbean countries have very low levels of PEM, and many have emphasized sugar production for export and chosen to pay to import much of their food. However, it should be pointed out that in environments with risky markets, joint promotion of both food and cash crops is required to achieve food security.

Developing countries should strive for integrated rural development combining sustainable agricultural development and the promotion of off-farm economic activities. Expanding agricultural efforts to increase and improve food production as well as to increase the income of rural families through greater production of cash crops is the job of most ministries of agriculture in developing countries.

Agricultural research in universities or in research stations is important to agricultural efforts. A good agricultural extension service can help farmers increase their productivity and make decisions about their farm practices. Agricultural research and extension, leading to higher levels of agricultural production, can have a major impact on nutrition, especially if improved production makes it easier for the poor to consume an adequate diet. Many textbooks examine how agriculture and food production are used to improve food intakes and nutritional status. They are essential reading for those who are interested in these aspects.

Local seasonal factors are very important influences on food supply. For example, rainfall patterns can give marked variations in food production within a year and between years. Food production can also be influenced by other factors such as pests, prices, availability of agricultural inputs and farmers' ability to procure them, political stability and peace. Climatic variations, especially rainfall (or its lack) and inclement weather, can dead to annual variations in food production. These variations may bring about complex food storage and management requirements. Seasonally high food prices may be tied to costs of storage and failure to manage public food stocks adequately.

Food storage limits and post-harvest losses due to insects, pests, moulds, bruising, high temperatures, etc. can seriously destabilize food supply. Yet even after production, harvest and storage are successfully accomplished, other factors can affect food supply. These include commercial food processing and industrialization; food marketing, including transport; policies related to importation and exportation of food, including food donated in multilateral or bilateral agreements; and external assistance and debt repayment.

Access to Food

Access to food, or food demand, is influenced by economic issues, physical infrastructure and consumer preferences.

Per caput incomes and food prices are important determinants of food demand. Since the poor are the most vulnerable to food deficits and malnutrition, policies that increase their purchasing power will provide them with the potential to improve their nutrition. Therefore, increased employment and better wages become components of policies and programmes to improve nutrition. In many poor countries the minority of the working population are wage-earners and the majority are self-employed in agriculture. About 65 percent of the population in developing countries of Asia and Africa and about 35 percent in Latin America live in rural areas and rely on agriculture, fishing, animal production and forestry for food as well as for income to purchase food and other necessities. Assistance to help this group of poor farmers and rural workers increase their incomes and food productivity will have an effect similar to that of increasing the wages of the urban poor.

Food prices affect both supply and demand. Lower prices give farmers less revenue for their produce. If prices drop too low, farmers may not produce or sell at all. However, lower prices represent an increase in the purchasing power of the consumer. Lowering the price of a common staple food such as maize or rice is equivalent to raising the income of all those who purchase this food. Similarly, raising the price (a more common occurrence) is equivalent to lowering the income of those who purchase it.

Governments have various mechanisms at their disposal to help satisfy the needs of both producers and consumers. One of these is subsidizing food prices: the price paid to the farmer for a sack of maize or rice is raised while market prices for consumers are maintained, with the government paying for the difference between the two. Food price subsidies may be disastrous for the economy but politically expedient for the government. They may help the poor to improve their nutrition.

Too often in the past, pricing policies and subsidies have been directed at foods consumed mainly by high-income groups and have thus had no beneficial effect on vulnerable groups. For example, price restrictions on meat, powdered milk or tinned baby foods or subsidies on beef or margarine would hardly benefit the poor at all, nor would they have important nutritional impact. Structural adjustment programmes put in place to mitigate severe economic crises often adversely affect the poor, particularly in the urban areas, through increased food prices. However, in many countries the majority of the rural poor are food producers, and structural adjustments may benefit them by raising their income from the sale of food produced and providing incentives to improve production efforts. By limiting inflation and reducing other macroeconomic distortions, structural adjustment programmes may benefit all population groups.

Food demand is also affected by consumer preferences, which can be shaped by cultural beliefs and practices or intra-household food allocation. An efficient infrastructure, including roads, railways, bridges and marketing facilities, is a determinant of the extent and success of food distribution to different segments of society. In the developing world and also in some industrialized countries, families living near food markets have a steady and easy access to cheaper foods and a more diversified diet, while people living far from markets usually have a rather narrow range of foods to choose from.

Household Food Security

Household food security is the ability of the household to secure enough food to provide for all the nutrient requirements of all members of the household. It is critical to link national food security and household food security, because availability of food supplies in adequate quantity and variety is a necessary but insufficient condition for ensuring adequate access by all households in need. Furthermore, having adequate overall food supplies in households is a necessary but insufficient condition for ensuring nutritionally adequate consumption by all individuals within households. Clearly, the overall availability of food in a country, community or household is no guarantee of its equitable consumption.

Components of Household food security

Household food security depends on a nutritionally adequate and safe food supply nationally, at the household level and for each individual; a fair degree of stability in the food availability to the household both during the year and from year to year; and access of each family member to sufficient food to meet nutritional requirements.

It is also important that the available food be both safe and of good quality. Attention to the food at every step of the food chain or food cycle is required to ensure its quality and safety. These steps include the cultivation of the food in the field (including protection against damage from pests or contamination with farm chemicals or pesticides); the harvesting, transport and storage of the food; processing and marketing; and finally the preparation and cooking of the food in the home and aspects of its consumption in the household. From the nutritionist's point of view, food losses and wastage along the chain are of great importance. However, important health concerns can also be raised if foods are not used correctly. An example is possible contamination, particularly from pesticides or other chemicals used to enhance production or to control pests such as insects, fungi, bacteria and viruses, or from natural toxins.

Food quality and safety are also affected by food hygiene, food handlers, people involved in food processing, those retailing the food and finally practices in the home. Certain codes and government inspections may help ensure some degree of safety, and education and knowledge of food hygiene by al] people will reduce the likelihood of contamination in the home. However, available facilities also influence food hygiene. Households that have poor facilities, no refrigeration, contaminated or inadequate water supplies or fuel shortages will find it more difficult to ensure food safety.

Another important aspect of food security is stability. The family or household must have the ability all year round to produce or procure the food its members require. The food must provide for all the family members' essential micronutrient and energy requirements, plus their wants, or desirable allowances, provided this does not lead to overconsumption. Of the greatest importance, especially when food or certain nutrients are available in marginal amounts, is proper distribution within the family to satisfy the special needs of children and females of childbearing age.

Incomes received from cash crops or wage earnings and prices paid for purchased items influence a rural population's food security. Inadequate landholdings, landlessness, sharecropping and other causes of poverty are all potent causes of family food insecurity. For the one-third of the population of developing countries who live in urban areas, much of the food obtained is purchased. The household food security of the urban poor depends on incomes, prices and the need to spend earnings on other essentials such as housing and transport. Their food security can be threatened by increased prices, job loss, income reduction, rent increases, larger numbers of dependent persons (more children, or relatives moving into the household) and other factors.

In both urban and rural areas the food must satisfy not only the energy needs but also the micronutrient needs of each household member. Therefore, the food consumed by each person must be varied and its quantity must be sufficient. If this is not the case, micronutrient deficiencies may occur.

Household Food Insecurity

Malnutrition may result from inadequate food, inadequate health or inadequate care. Inadequate food, be it due to food shortages or to inappropriate consumer behaviour or intrahousehold distribution, is termed food insecurity.

Food insecurity at the household or individual level may be transitory, or short-term, because of a particular event of short duration. In these circumstances it results from a temporarily limited access to food. Chronic food insecurity is long-term, may have a more marked impact and may be more difficult to control. The intensity of either short-term or long-term food insecurity is also important. Food insecurity occurs in mild, moderate and severe forms, just as PEM does. The level of food insecurity may be related to the relative availability of food.

A "shock" often precipitates household food insecurity. The shock can aggravate poverty (suddenly making a poor family very poor) or adversely influence food production (suddenly threatening farm food availability). There are many different kinds of shock, for example, serious illness, which may result in loss of income in an urban family or reduced agricultural production in a farm family; loss of a rural or urban job; farm production crises, such as failure of the rains; or a plague of locusts or some other agricultural catastrophe. Any crisis that has an adverse impact on the livelihood of the family may also result in household food insecurity.

Another important determinant of food insecurity is gender discrimination. Subordination of women in society, their overburdening and the greater difficulties faced by female-headed households contribute to food insecurity.

Food Policies in a Development Context

Clearly, development strategies and interventions pursued by developed and developing nations have an impact on nutrition. For this impact to be positive, developed and developing countries must decide what "development" really means.

Too often in the past, development has been associated with industrialization and measured by the productive capacity and the material output of a country. Indicators of development were gross national product (GNP) or mean per caput incomes. Economists tended to view improved nutrition and health as welfare questions. However, it is now clear that economic development does not benefit everyone equally. The poor have often been bypassed, and improvements in the quality of life of most low-income families in many countries have not kept pace with the improvements in national economic figures. The purpose and the intended beneficiaries of economic development should be examined before the interventions begin. If development plans do not encompass improved health and better nutrition for people, then their worth must be seriously questioned.

Nutrition-positive development projects are those that will benefit a large segment of the population, help reduce inequalities in income distribution and be likely to improve the nutrition, health and quality of life of those currently deprived. Labour-intensive projects are often preferable to capital-intensive ones, and support for small farmers may be more useful in regard to nutrition than assistance for large estates. Small farmers and especially women farmers are the most disadvantaged and require the most help. They are also the ones who receive the least assistance, in terms of both extension services and access to credit. In many countries, too little of the national budget is devoted to support for agriculture, which is essential for social and economic development and for nutritional well-being.

Food policy should make marketing as logical, simple and well-organized as possible, with a minimum involvement of intermediaries, to help ensure that the producer gets a fair return for his or her produce and that the consumer pays the lowest reasonable price for his or her food. Cooperatives are one form of marketing that may benefit both producer and consumer.

Recently, both adequate food and good nutrition have been declared basic human rights. Good nutrition goes beyond food rights, including also adequate care and adequate health. It has been suggested that household food security should be examined as part and parcel of a broader food and nutrition system. Food factors included in the system are food production and some of the influences on it; the transport system; the market and its relationship to exchange and storage; and finally household food availability and access. Most "food systems" do not give consideration to the health causes of malnutrition such as infections including diarrhoea and intestinal worms. They also do not include caring factors that may influence nutritional status, such as breastfeeding, weaning and psycho-social stimulation.

Nutrition and Infection, Health and Disease

The interaction or synergism of malnutrition and infection is the leading cause of morbidity and mortality in children in most countries in Africa, Asia and Latin America. Viral, bacterial and parasitic infections tend to be prevalent, and all can have a negative impact on the nutritional status of children and adults. The situation was similar in North America and Europe from about 1900 to 1925; common infectious diseases had an impact on nutrition and produced high case fatality rates.

The synergistic relationship between malnutrition and infectious diseases is now well accepted and has been conclusively demonstrated in animal experiments. The simultaneous presence of both malnutrition and infection results in an interaction that has more serious consequences for the host than the additive effect would be if the two worked independently. Infections make malnutrition worse and poor nutrition increases the severity of infectious diseases.

Effects of Malnutrition on Infection

The immune system

The human body has the ability to resist almost all types of organisms or toxins that tend to damage the tissues and organs. This capacity is called immunity. Much of the immunity is caused by a special immune system that forms antibodies and sensitized lymphocytes which attack and destroy the specific organisms or toxins. This type of immunity is called acquired immunity. An additional portion of the immunity results from the general processes of the body; this is called innate immunity

Innate immunity is due to:

— resistance of the skin to invasion by organisms;

— phagocytosis of bacteria and other invaders by white blood cells and cells of the tissue macrophage system;

— destruction by the acid secretions of the stomach and by the digestive enzymes of organisms swallowed into the stomach;

— the presence in the blood of certain chemical compounds that attach to the foreign organisms or toxins and destroy them.

There are two basic but closely allied types of acquired immunity. In one of these the body develops circulating antibodies, which are globulin molecules that are capable of attacking the invading agents and destroying them. This type of immunity is called humoral immunity. Antibodies circulate in the blood and may remain there for a long time, so that a second infection with the same organism is immediately controlled. This is the basis for some forms of immunization, which are designed to stimulate antibody production.

The second type of acquired immunity is achieved through the formation of large numbers of highly specialized lymphocytes which are specifically sensitized against the invading foreign agents. These sensitized lymphocytes have the ability to attach to the foreign agents and to destroy them. This type of immunity is called cellular immunity. It is a highly complex system involving many different body organs (such as the spleen, thymus, lymph system and bone marrow) and also body fluids, particularly blood and its constituents and lymph.

The study of the complex system of immunity is termed immunology.

Effects of Malnutrition on Resistance to Infection

A considerable amount of literature, documenting studies both in experimental animals and in people, demonstrates that dietary deficiency diseases may reduce the body's resistance to infections and adversely affect the immune system.

Some of the normal defence mechanisms of the body are impaired and do not function properly in the malnourished subject. For example, children with kwashiorkor were shown to be unable to form antibodies to either typhoid vaccine or diphtheria toxoid; their capacity to do so was restored after protein therapy. Similarly, children with protein malnutrition have an impaired antibody response to inoculation with yellow fever vaccine. An inhibition of the agglutinating response to cholera antigen has been reported in children with kwashiorkor and nutritional marasmus. These studies provide a fairly clear indication that the malnourished body has a reduced ability to defend itself against infection.

Another defence mechanism that has been studied in relation to nutrition is that of leucocytosis (increased production of white blood cells) and phagocytic activity (destruction of bacteria by white corpuscles). Children with kwashiorkor show a lower than normal leucocyte response in the presence of an infection. Perhaps of greater importance is the reduced phagocytic efficiency in malnourished subjects of the polymorphonuclear leucocytes that are part of the fight against invading bacteria. When malnutrition is present, these cells appear to have a defect in their intracellular bactericidal (bacteria-destroying) capacity.

Although malnourished children frequently have increased immunoglobulin levels (presumably related to concurrent infections), they also may have depressed cell-mediated immunity. In a recent study, the extent of this depression was directly related to the severity of the protein-energy malnutrition (PEM). Serum transferrin levels are also low in those with severe PEM, and they often take considerable time to return to normal even after proper dietary treatment.

A quite different kind of interaction of nutrition and infection is seen in the effect of some deficiency diseases on the integrity of the tissues. Reduction in the integrity of certain epithelial surfaces, notably the skin and mucous membranes, decreases resistance to invasion and makes an easy avenue of entry for pathogenic organisms. Examples of this effect are cheilosis and angular stomatitis in riboflavin deficiency, bleeding gums and capillary fragility in vitamin C deficiency, flaky-paint dermatosis and atrophic intestinal changes in severe protein deficiency and serious eye lesions in vitamin A deficiency.

Effects of Infection on Nutritional Status

Infection affects nutritional status in several ways. Perhaps the most important of these is that bacterial and some other infections lead to an increased loss of nitrogen from the body. This repercussion was first demonstrated in serious infections such as typhoid fever, but it has subsequently been shown in much milder infections such as otitis media, tonsillitis, chicken pox and abscesses.

Nitrogen is lost by several mechanisms. The principal one is probably increased breakdown of tissue protein and mobilization of amino acids, especially from the muscles. The nitrogen is excreted in the urine and is evidence of a depletion of body protein from muscles.

Full recovery is dependent upon the restoration of these amino acids to the tissues once the infection is overcome. This requires increased intake of protein, above maintenance levels, in the post-infection period. In children whose diet is marginal in protein content, or those who are already protein depleted, growth will be retarded during and after infections. In developing countries, children from poor families suffer from many infections in quick succession during the post-weaning period, and they often have multiple infections.

Anorexia or loss of appetite is another factor in the relationship between infection and nutrition. Infections, especially if accompanied by a fever, often lead to loss of appetite and therefore to reduced food intake. Some infectious diseases commonly cause vomiting, with the same result. In many societies mothers and often medical attendants as well consider it desirable to withhold food or to place the child with an infection on a liquid diet. Such a diet may consist of rice water, very dilute soups, water alone or some other fluid with a low calorie density and usually deficient in protein and other essential nutrients. The old adage of "starve a fever" is of doubtful validity, and this practice may have serious consequences for the child whose nutritional status is already precarious.

The traditional treatment of diarrhoea in some communities is to prescribe a purgative or enema. The gastro-enteritis may already have resulted in reduced absorption of nutrients from food, and the treatment may further aggravate this situation.

These are all examples of how illnesses such as measles, upper respiratory infections and gastro-intestinal infections may contribute to the development of malnutrition.

Parasitic Infections

Parasitic infections, particularly intestinal helminthic infections, are extremely prevalent and are increasingly being shown to have an adverse effect on nutritional status, especially in those heavily infected. Hookworms *(Ancylostoma duodenale* and *Necator americanus)* infect over µ00 million people, mainly the poor in tropical and subtropical countries. They used to cause a prevalent debilitating disease in the southern United States. Hookworms cause intestinal blood loss, and although it appears that most of the protein in the lost blood is absorbed lower down in the intestinal tract, there is considerable loss of iron.

Hookworm disease is a major cause of iron deficiency anaemia in many countries. The extent of the loss of blood and iron in hookworm infections has been studied: daily faecal blood loss per hookworm (*N. americanus*) was reported to be 0.031 ± 0.015 ml. It was estimated that about 350 hookworms in the intestine cause a daily loss of 10 ml of blood, or 2 mg of iron. Infection densities much higher than this are not uncommon. In Venezuela, where much of this work was done, iron losses greater than 3 mg per day often resulted in anaemia in adult males, and losses of half this amount frequently produced anaemia in women of child-bearing age and in young children.

Worldwide, roundworm (*Ascaris lumbricoides*) is among the most prevalent of intestinal parasites. It is estimated that 1 200 million people in the world (one-quarter of the world's population) harbour roundworms. The roundworm is large (15 to 30 cm long), so its own metabolic needs must be considerable. High parasite densities, particularly in children, are common in environments where sanitation is poor. Complications of ascariasis can develop, including intestinal obstruction or the presence of worms in aberrant sites such as the common bile duct. In some countries ascarids are a cause of surgical emergencies in children, and many with obstruction die. In the majority of children, however, when malnutrition is prevalent, deworming improves child growth.

Trichuris trichiura or whipworm inhabits the large intestine and infects about 600 million people worldwide. The worms are small and, in heavily infected children, may cause diarrhoea and abdominal pain.

Many children living in poor sanitary conditions are infected with several parasitic infections at the same time. In areas where infection with these three parasites is common and where malnutrition is prevalent, deworming of children leads to an improvement in growth, a reduction in the extent of malnutrition and an increase in appetite. It also positively influences physical fitness and perhaps psychological development.

Bilharzia or schistosomiasis infections are prevalent in some countries. They also contribute to poor nutrition, poor appetite and poor growth. The three organisms that cause schistosomiasis *(Schistosoma haematobium, Schistosoma mansoni* and *Schistosoma japonicum*) are flukes, rather than ordinary worms.

Somewhat less is known about the relationship between intestinal protozoa! diseases and nutrition, but amoebas, causing serious dysentery and liver abscess, are highly pathogenic organisms, and infection with *Giardia lamblia* may cause malabsorption and abdominal pain.

The fish tapeworm *(Diphyllobothrium latum*) has an avidity for vitamin B_{12} and can deprive its host of this vitamin, with megaloblastic anaemia resulting. The fish tapeworm is common in people in only limited geographic areas, mainly in temperate areas and where undercooked fish is frequently consumed.

In many northern industrialized countries, farm animals and domestic pets such as dogs and cats are dewormed routinely. Much evidence suggests that pigs grow better when they regularly receive anthelmintics. Now that highly effective, relatively inexpensive and safe broad-spectrum anthelmintics such as albendazole and mebendazole are available, routine mass deworming should be introduced where parasitic infections are prevalent in humans and where PEM and anaemia are common. Similarly, routine efforts to treat children with schistosomiasis using metrifonate or praziquantel seem highly desirable both to rid children of potential serious pathology and to improve their nutritional status. More attention needs to be given to population-based chemotherapy for these infections along with intensification of public health and other measures to reduce their transmission, including improved sanitation and water supplies. Such efforts would improve the health and nutritional status of millions of the world's children.

Effects of Diarrhoea

Many studies have indicated that gastrointestinal infections, and especially diarrhoea, are very important in precipitating serious PEM. Diarrhoea is common in, and often lethal to, the young child. In breastfed infants there is often some protection during the first months of life, so diarrhoea is often a feature of the weaning process. Weanling diarrhoea is extraordinarily prevalent in poor communities throughout the world, both in tropical and temperate zones. The organism responsible varies and often cannot be identified. Diarrhoea was a major cause of mortality in children in industrialized countries up to the beginning of the twentieth century.

Several studies have shown that admissions of cases of malnutrition are greatly increased during the season when diarrhoea is most common. For example, in a report from the Islamic Republic of Iran, more than twice as many cases of PEM were admitted in the warm summer than in the cold winter. The incidence of diarrhoeal disease followed the same pattern.

Hospital and community studies indicate that cases of xerophthalmia and keratomalacia are frequently precipitated by gastro-enteritis, as well as by other infectious diseases such as measles and chicken pox. Xerophthalmia is the major cause of blindness in several Asian countries; it is also prevalent in certain parts of Africa, Latin America and the Near East.

Intestinal parasites may contribute to diarrhoea and to poor vitamin A status. The exact mechanism of this relationship has not been proved, but it is likely that many infections reduce vitamin A absorption and that some result in decreased consumption of foods containing vitamin A and carotene.

Diarrhoea can be fatal, usually because it can lead to severe dehydration. Diarrhoea, and the complication of dehydration, may be said to be a form of malnutrition. Dehydration is a "deficiency" in the body of water and mineral electrolytes, and providing adequate quantities of these cures the deficiency. The term "fluid electrolyte malnutrition" (FEM) has been coined for this condition. Provision of water and adequate minerals in home-prepared food, breastfeeding or administration of oral rehydration fluids is now the accepted treatment. Although these are forms of therapy or treatment, they are really refeeding and replenishment. However, prevention requires measures and interventions to reduce infections, poverty and malnutrition. These are essential if countries are to reduce the incidence of diarrhoea.

Fatality Rates for Measles and Other Infectious Diseases

A dramatic illustration of the effect of malnutrition on infection is seen in the fatality rates for common childhood diseases such as measles. Measles is a severe disease with a case fatality rate of about 15 percent in many poor countries because the young children who develop it have poor nutritional status, lowered resistance and poor health. In Mexico the fatality rate for measles has been reported to be 180 times higher than that in the United States; in Guatemala, 268 times higher; and in Ecuador, 480 times higher. The decline in case fatality rates of measles in North America, Europe and other industrialized countries has been dramatic over the last century.

Differences in the clinical severity and the fatality rates of measles in developed and developing countries are due not to differences in virus virulence but to differences in the hosts' nutritional status. For example, during a measles epidemic in the United Republic of Tanzania that was causing considerable mortality among the children of poorer families, it was observed that fatalities from the disease were extremely uncommon in the children of families of moderate income, such as those of hospital employees. Measles is also related to vitamin A deficiency. It has been shown that providing vitamin A supplements to children with measles who have poor vitamin A status greatly reduces case fatality rates.

Immunization against measles is proving very effective, and in many countries measles incidence has been markedly reduced.

Other common infectious diseases such as whooping cough, diarrhoea and upper respiratory infections also have much more serious consequences in malnourished children than in those who are well nourished. Mortality statistics from most developing countries show that such communicable diseases are the major causes of death. It was observed in several African countries at the end of the Sahel famine that very few children were dying of starvation or malnutrition, but that deaths from measles, respiratory infections and other infectious diseases were still very much above pre-famine levels. It is clear that many, perhaps the majority, of these deaths were due to malnutrition. This may seem a moot point for a grieving parent, but for the policy planner and the public health official it is important to know to what extent morbidity and mortality rates are due to or related to undernutrition.

An inter-American investigation of mortality in childhood showed that of 35 000 deaths of children under five years of age in ten countries, in 57 percent of the cases malnutrition was either the underlying or an associated cause of death. Nutritional deficiency was the most serious health problem uncovered, and it was frequently associated with common infectious diseases.

HIV Infection and AIDS

Perhaps no disease has a more dramatic and obvious effect on nutritional status than acquired immunodeficiency syndrome (AIDS), the disease caused by the human immunodeficiency virus (HIV). In Uganda for many years the disease was called "slim disease" because extreme thinness was the main visible manifestation of the disease. Although the mechanisms by which AIDS leads to severe malnutrition have not been proven, there is no doubt that the disease and its associated opportunistic infections cause marked anorexia, diarrhoea and malabsorption as well as increased nitrogen losses. Some of the infections and conditions that are part of the AIDS complex of diseases were known to affect nutritional status long before the HIV virus was identified: tuberculosis has for many decades been associated with cachexia and weight loss, and malignancies such as sarcoma have long been known to result in wasting as they advance.

Chronic Diseases and Old Age

There is a relationship between certain chronic diseases and immune response. It has also been clearly shown that in old age immunologic response is reduced, and undernutrition worsens this decline. The association of diabetes with infections is well known, and it is clear that in diabetes there is often impaired cellular response. Other diseases, for example several cancers, may also be related to lowered immune response.

Intervention Studies

There have been relatively few well-controlled intervention studies to demonstrate either the effects of improved diets on infection or the nutritional effects of control of infectious diseases. Research in the village of Candelaria in Colombia showed that diarrhoea declined sharply as a result of supplementary feeding of children. A similar study in a Guatemalan village illustrated a significant decline in morbidity and mortality from certain common illnesses following the introduction of a nutritious daily supplement for preschool children.A classic study conducted in Narangwal in the Punjab region of India demonstrated the value of combining nutritional care and health care in one programme. Children were divided into four groups. One group was given dietary supplements, one group was given health care, one group received both the supplements and the health care, and the fourth group served as control. As far as nutritional status and certain other health parameters were concerned, the combined treatment gave the best results. Nutritional supplementation alone also had a major impact. In comparison with the control group, there was no improvement in the nutritional status of the group that received only medical care but no dietary supplements.

Nutrition, Infection and National Development

Clearly, the effects of nutritional status on infections and of infections on malnutrition signify a very important relationship. The majority of children in most developing countries suffer from malnutrition at some time in their first five years of life. The problems of infection and malnutrition are closely interrelated, yet programmes to control communicable diseases and to improve nutrition tend to be introduced quite independently. It would be much more efficient and effective if the twin problems were attacked together.

Success in improving the health and reducing the mortality of children is dependent both on control of infectious diseases and on improvements in the children's food intake and care. There is increasing evidence to suggest that parents are more willing to control their family size when the chances are good that most children born will survive into adulthood. Consideration also needs to be given to providing a stimulating environment for the growing child.

The situation in the major industrial cities of Europe and North America a century ago was comparable to that in the poorest developing countries today. In New York City in the summer months of 1892, the infant mortality rate was 340 per 1 000, and diarrhoea. accounted for half these deaths. Improvements in nutrition, through the use of milk stations, for example, and a reduction in infectious disease served to lower these mortality rates by half in a period of less than 25 years. In the United Kingdom at the beginning of the twentieth century, rickets, combined with infectious diseases, took a heavy toll in the insanitary, smoky slums of the industrial cities, and measles was very often fatal among children of poor families, presumably because of poor nutrition.

Malnutrition and infections combine to pose an enormous hazard to the health of the majority of the world's population who live in poverty. This ever-present hazard particularly threatens children under five years of age. Many of the children who suffer from both malnutrition and a series of infections succumb and die. They are continually replaced in answer to parents' strong desire and often real need to have surviving children. The children who live beyond five years of age are not mainly those who have escaped malnutrition or infectious diseases, but those who have survived. Seldom are they left without the permanent sequelae or scars of their early health experiences. They are often retarded in their physical, psychological or behavioural development, and they may have other abnormalities that contribute to a less than optimal ability to function as adults and possibly to a shortened life expectation. Other factors influencing the development of these children include a lack of environmental stimulation and a host of other deprivations related to poverty.

The challenge to health workers, development economists, governments and international agencies is how best to reduce the morbidity, mortality and permanent sequelae that result from the synergism of malnutrition and infection. The politicians must be persuaded that attention to these problems is not only highly desirable but politically advantageous.

The control of infectious diseases and projects aimed at providing more and better food for people are fully justified and important components of a development plan. By themselves they may contribute to increased productivity and better lives. An improved infant or toddler mortality rate, a lowered disease incidence and a better-nourished population are probably better indicators of development shall national averages of telephones or automobiles per 1 000 families, or even than dollars or pesos per caput. Efforts for the control of infectious diseases and the improvement of nutrition both deserve a high priority in development plans and in international or bilateral assistance to low-income countries. They should be undertaken together because they will be mutually reinforcing and more economical if provided in a coordinated manner rather than separately. An allied issue is the need to provide a stimulating environment for the growing child.

Historical and epidemiological evidence suggests that reductions in infant and child mortality and improvements in health and nutritional status may be prerequisites to successful family planning efforts. Birth spacing deserves a high priority, especially where women are already overworked and undernurtured. Parents in all countries should receive assistance to help them achieve their desired family size.

Alarming as the situation of children's malnutrition and infection is, there is a general tendency to overlook the significance of these conditions in adults. Weakness, lethargy, absenteeism, poor productivity and stress can all have social and economic costs for individuals, families and communities.

There seems to be unassailable logic in recommending coordinated programmes that have three objectives: to control infectious disease, to improve nutrition and to make family planning services widely available. These three types of endeavour may themselves be synergistic.

References

Berg, A.(1987). *Malnutrition. What can be done? Lessons from the World Bank experience.* Baltimore, Maryland, USA, Johns Hopkins University Press.

Brun, T.A. & Latham, M.C. (1990). *Maldevelopment and malnutrition.* World Food Issues, Vol. 2. Ithaca, New York, USA, Cornell University, Program in International Agriculture.

UN ACC/SCN.(1989). *Malnutrition and infection. A review,* by A. Tomkin & F. Watson. Geneva, Switzerland.

Waterlow, J.C. (1992). *Protein energy malnutrition.* London, UK, Edward Arnold.

World Bank.(1994). *Enriching lives. Overcoming vitamin and mineral malnutrition in developing countries.* Washington, DC, USA.

8

Socio-cultural Aspects of Human Nutrition

Social factors and cultural practices in most countries have a very great influence on what people eat, on how they prepare food, on their feeding practices and on the foods they prefer. Nonetheless, cultural food practices are very rarely the main, or even an important, cause of malnutrition. On the contrary, many practices are specifically designed to protect and promote health; providing women with rich, energy-dense foods during the first months following childbirth is an example. It is true, however, that some traditional food practices and taboos in some societies may contribute to nutritional deficiencies among particular groups of the population. Nutritionists need to have a knowledge of the food habits and practices of the communities in which they work so that they can help to reinforce the positive habits as well as strive to change any negative ones.

Food Habits

All people have their likes and dislikes and their beliefs about food, and many people are conservative in their food habits. They tend to like what their mothers cooked for them when they were young, the foods that are served on festive occasions or those eaten with friends and family away from home during their childhood. The foods that adults ate without a second thought in childhood are seldom totally disagreeable to them in later life.

What one society regards as normal or even highly desirable, however, another society may consider revolting or totally inedible. Animal milk is commonly consumed and liked by many people in Asia, Africa, Europe and the Americas, but in China it is rarely taken. Lobsters, crabs and shrimps are considered delicacies and prized foods by many people in Europe and North America, but are revolting to many people in Africa and Asia, especially those who live far from the sea. The French eat horse meat; the English generally do not. Many people will delightedly consume the flesh of monkeys, snakes, dogs and rats or will eat certain insects, yet many others find these foods most unappealing. Religion may have an important role in forbidding the consumption of certain foods. For example, neither the Muslim nor the Jewish peoples consume pork, and Hindus do not eat beef and are frequently vegetarians.

Food habits differ most widely in regard to which foods of animal origin are liked, disliked, eaten or not eaten in a society. The foods in question comprise many of those that are rich in good-quality protein and that contain haem iron, both of which are important nutrients. People who do not consume these foods are deprived of the opportunity of obtaining these nutrients easily. On the other hand, those who overconsume animal flesh, some seafoods, eggs and other foods of animal origin will have undesirable amounts of saturated fat and cholesterol in the diet. Balanced consumption is the key.

Relatively few people or societies have strong negative feelings about consuming cereals, roots, legumes, vegetables or fruit. They may have strong preferences and likes, but most maize-eating people are also willing to eat rice, and most rice-eating people will eat wheat products.

It is often stated that food habits seldom or never change and are difficult to change. This is not true; in many countries the current staple foods are not the same as those eaten even a century ago. Food habits and customs do change, and they are influenced in many different ways. Maize and cassava are not indigenous to Africa, yet they are now major food staples in many African countries. Potatoes originated in the Americas and later became an important food in Ireland.

Food preferences are not made and abolished by whims and fancies, of course. More often the adjustments are generated by social and economic changes that take place throughout the community or society. The issue is often not what foods are eaten but rather how much of each food is eaten and how the consumption is distributed within the society or within the family.

The tendency of many wage-earners to spend almost all their wages within a few days of receiving them often results in a family diet of varying nutritive value. The family eats much better just after one payday than just before the next. Wages are often paid monthly, and there seems little doubt that a change to weekly payment of wages would improve the diet of wage-earners and their families.

The person who controls the family finances influences (intentionally or unintentionally) both the family diet and the food fed to children. In general, when mothers, rather than just fathers, have some control over finances, the family diet is likely to be better. When the mother has little control over family funds, dietary arrangements may become haphazard or even dangerous.

Nutrition education has been an important influence on food habits, and not always a positive one. Fortunately, the days are long gone when nutritionists promoted costly protein-rich foods to eople who couldn't possibly afford them. Unfortunately, the tendency to single out foods or nutrients either to promote or to prohibit has not yet gone, nor has the tendency to try to teach by creating fear and taking the enjoyment out of eating. However, change always comes slowly and old habits die hard; people who were taught in these old ways are still responsible for feeding themselves and their families, and they may find it hard to change again.

Nutritional Advantages of Traditional Food Habits

The traditional diets of most societies in developing countries are good. Usually only minor changes are needed to enable them to satisfy the nutrient requirements of all members of the family.

Eating certain protein-rich foods such as insects, snakes, baboons, mongooses, dogs, cats, unusual seafoods and snails is definitely beneficial. Another habit that is good nutritionally is the consumption of animal blood. Some African tribes puncture the vein of a cow, draw off a calabash of blood, arrest the bleeding and consume the blood, usually after mixing it with milk. Blood is a rich food, and mixed with milk it is highly nutritious.

A custom frequently found among pastoral and other peoples is the drinking of soured or curdled milk, rather than fresh. The souring of milk has little effect on its nutritive value but often substantially reduces the number of pathogenic organisms present. In communities where milking is not hygienically performed and where the containers into which the milk goes are likely to be contaminated, it is safer to drink sour rather than fresh milk. Boiled milk would be safer still.

Many societies, for example in Indonesia and in parts of Africa, partly ferment foods before consumption. Fermentation may both improve the nutritional quality and reduce bacterial contamination of the food.

The traditional use of certain dark green leaves among rural peoples is another beneficial practice and should be encouraged. These leaves are rich sources of carotene, ascorbic acid, iron and calcium; they also contain useful quantities of protein. Non-cultivated or wild dark green leaves such as amaranth leaves as well as those from cultivated food crops such as pumpkin, sweet potato and cassava are much richer in vitamins than pale, leafy vegetables of European origin such as cabbage and lettuce. Well-meaning expatriate horticulturists in Africa have too often tried to get villagers to cultivate such European vegetables rather shall their traditional vegetables.

Many wild fruits are rich in vitamin C; an example is the pulp within the pod of the frequently consumed baobab.

Traditional grain preparation methods produce a more nutritious product than does elaborate machine milling.

Some communities sprout legume seeds prior to cooking, which enhances their nutritive value, as does the soaking of whole-grain cereals before their processing into local beers and some non-alcoholic beverages. These seeds and grains usually have a high vitamin B content. Finally, it cannot be stressed too strongly that the traditional method of infant feeding-from the breast - is nutritionally far superior to bottle-feeding.

Food Taboos

A number of food habits and practices are poor from a nutritional point of view. Some practices result from traditional views about food that are liable to change under the influence of neighbouring peoples, travel, education, etc. Other food practices are governed by definite taboos.

A taboo may be followed by a whole national group or tribe, by part of a tribe or by certain groups in the society. Within the society, different food customs may be practiced only by women or children, or by pregnant women or female children. In certain cases traditional food customs are practiced by a particular age group, and in other instances a taboo may be linked with an occupation such as hunting. At other times or in other individuals a taboo may be imposed because of some particular event such as an illness or an initiation ceremony.

Although these matters border on the realm of anthropology, it is important for a nutritionist to be familiar with the food customs of people in order to be able to improve their nutritional status through nutrition education or other means. Moreover, it is evident that anthropology and sociology are important to the nutrition worker who is either investigating or trying to improve the nutritional status of any community.

Some customs and taboos have known origins, and many are logical, although the original reasons may no longer be known. The custom may have become part of the religion of the people involved. For example, the Jewish taboo against pork was probably introduced to eliminate the prevalent pork tapeworm, which was thought to be sapping the strength of the Jewish people. Even though 2 000 years later it is now possible to eat pork safely, Jews still do not eat pork. Muslims share this view about pork. In neither case is this a nutritionally damaging taboo.

Many taboos concern the consumption of protein-rich animal foods, often by those groups of the community most in need of protein. A common taboo in Africa against the consumption of eggs is rapidly disappearing. This taboo usually applies to females, who are said to become sterile if they eat eggs. The psychological connection between human fertility and the egg is obvious. In other places the custom applies to children, perhaps to discourage them from stealing the eggs of setting hens, which would endanger the survival of poultry. Other customs, again often affecting women and children, concern fish. These customs may amount to a full taboo, although people not used to fish often dislike it merely because they find its smell distasteful or its appearance "snake-like". Many cultures have strong views about the consumption of milk or milk products.

The customs that prohibit consumption of certain nutritionally valuable foods may not have an important overall nutritional impact, particularly if only one or two food items are affected. Some societies, however, forbid such a wide range of foods to women during pregnancy that it is difficult for them to obtain a balanced diet.

Many of the nutritionally undesirable taboos that existed a quarter of a century ago have weakened or disappeared as a result of education, mixing of people from different societies and travel. Of those that remain, some food habits may seem illogical and their origins obscure, but it is not advisable for outsiders to try to alter ancient food habits without looking very closely into their origins. Moreover, it makes no sense to attempt to alter a habit that does not negatively affect nutritional status.

Nutritionally bad habits, like all other habits, are best changed by the people who have them. In this regard, influential local people, with the welfare of their fellows at heart, may join nutritionists and become part of an important alliance pledged to eradicating malnutrition. A speech by the president or a cabinet minister, the sight of a respected tribe leader eating some forbidden food and coming to no harm or the return to the village of educated and enlightened local people will prove much more effective than the preaching or goading of an outsider.

Changing Food Habits

In some parts of the world the staple foods are changing or have changed. Maize, cassava and potatoes, now grown in large amounts in Africa, originated outside the continent. Since none of these foods were eaten in Africa a few hundred years ago, it is clear that the food habits of millions of people have changed. Vast numbers of people in Africa have abandoned yams and millet for maize and cassava, just as many in Europe abandoned oats, barley and rye for wheat and potatoes. Food habits are still changing rapidly. The difficulty, of course, lies in trying to guide and foster desirable changes and to slow down undesirable ones.

It is often difficult to fathom what factors have been most important in stimulating or influencing changes in food habits. The rapid increase in bread consumption in many African, Latin American and Asian countries where wheat is not the staple food is understandable. It is at least in part a labour-saving phenomenon; bread is one of the first "convenience" foods to have become available. Before leaving home to go to work one can eat some slices of bread instead of the traditional breakfast of porridge, which requires preparation time and is unpleasant cold. Bread can be carried in the pocket and eaten during a break in the working day, or when travelling.

In most of the world the traditional main staple food has remained constant, irrespective of urbanization, modernization or even westernization. Thus in much of Asia rice remains the preferred staple food in rural and urban areas. Some people in Africa, such as the Buganda in Uganda and the Wachagga in the United Republic of Tanzania, continue to have a preference for plantains as their staple food. Maize based products such as tortillas remain important in the diets of most Mexicans and many in Central America.

Changes in food habits are not just accidental, of course; they can be deliberately initiated. At community and family level, school-age children can be important agents for change. They are still forming their tastes and developing their preferences. If they are introduced to a new food they will often readily accept it and like it. School meals may usefully introduce new foods to children and thus influence food habits. This widening of food experience in childhood is extremely important. Children may influence the immediate family and later their own children to eat new, highly nutritious foods.

Harmful New Habits

Not all change is desirable, of course, and not all new food habits are good. Less attention has been given to the question of other baby foods that have been marketed and much promoted and advertised in developing countries. Locally available complementary or weaning foods, home-produced and traditionally fed, are often as or more nutritious than the manufactured baby foods, and then are always much cheaper. They are usually introduced gradually while breastfeeding continues well into the second year and beyond. Manufactured baby foods should only be promoted to those who are unable or unwilling to continue breastfeeding. They are safe and nutritionally adequate when prepared hygienically and in the right dilution. They are convenient for those who can afford to purchase them. However, such manufactured foods are expensive compared with local foods, and for most families in developing countries, other than the very affluent, they may be a waste of money. For families who already have too little money to spend on food and other essentials, these foods are a very expensive way of buying the nutrients that they are advertised to contain.

Another particularly misleading type of advertising relates to the glucose products said to provide "instant energy". Energy is present in large amounts in nearly all the cheapest foods. Similarly, drinks advertised as "rich in vitamin C" are usually unnecessary, since few children suffer from vitamin C deficiency. Vitamin C can be obtained just as well from fruits such as guavas, mangoes and citrus, or from a range of vegetables.

The so-called protein-rich weaning foods are also much advertised. These arc nutritionally good products, but they cost much more shall protein-rich foods available in the market such as beaus, groundnuts or dried fish, meat, eggs or milk. It usually costs much more to provide 100 g of protein from these commercially advertised products than, for example, from beans bought in the local market. The essential question is how a mother could best improve her child's diet if she had a little extra to spend. The answer would seldom be a manufactured baby food.

In some countries the staple food has remained unaltered, but the form in which it is preferred may have changed over the years. The rapid spread and popularity of highly milled rice in Asia had disastrous consequences and led to a high prevalence of beriberi, with much morbidity and many deaths. In many parts of the world highly milled cereals have replaced traditionally lightly milled and more nutritious wheat, rice and maize. In the United Kingdom and the Russian Federation, white bread has replaced brown or whole-grain breads, and in East Africa highly milled maize meal is often purchased and has replaced lightly milled maize flour. Urbanization, modernization and sophistication have often led to diets in which a greater percentage of energy intake comes from sugar and fats, and to increased consumption of salt. All Of these are generally undesirable changes from a nutritional standpoint.

Influencing Change for the Better

What can health workers or nutritionists in a community do about food habits, old and new? They can:

— protect, support and help preserve the many excellent existing food habits that are nutritionally valuable;

— respect the knowledge and customs of the people in the community in which they work;

— set good examples in their own households by adopting good food habits;

— influence respected local leaders to state publicly that they themselves have dropped undesirable food taboos, and arrange for them, when occasion arises, to eat "forbidden" foods in public;

— persuade people not to abandon good food habits under the influence of "sophisticates" back from the city who may try to discourage rural dwellers from eating nutritious traditional foods such as locusts or lake flies or to encourage the consumption and production of European-type vegetables in place of better traditional ones;

— explain the disadvantages of highly refined cereal flours if they have become popular in the area, and advocate the consumption of a range of cereals in the local diet;

— discourage poorer families from purchasing manufactured baby foods, and encourage the use of locally available complementary foods;

— issue informational material to help stop the spread of bottle-feeding and the unnecessary purchase of expensive baby foods;

— strive, through civil service or local authority organizations, for the introduction of the payment of weekly wages instead of monthly wages to employees, and influence labour and trade union leaders to do the same;

— take steps to introduce good feeding practices in the local schools and other institutions.

Population, Food and Nutrition

Many thinkers in the world and many who work in the development field believe that the world's population size and increase is its greatest problem and humanity's gravest threat. Clearly the ratio of the number of people to the amount of food available has an impact on nutrition, but how are the two caused to interact? Late in the eighteenth century the British political economist Thomas Malthus grimly speculated that population growth could soon outstrip food production and supply. Close to the end of the twentieth century this has not yet happened, but malnutrition is widespread.

Population Growth

World population is increasing at an alarming rate. Unless the rate of increase is slowed down in the next few decades, the world will face extremely serious problems. The world population was around 250 million people 2 000 years ago. After taking 16 centuries to double to 500 million, it then doubled in two and a half centuries to reach 1000 million in 1850, and it doubled again in one century to reach 2 000 million people in 1950. Now the population of the world is doubling every 35 years; it reached 5 000 million before 1990.

Population pressure is most marked and is having a major impact in Asian countries such as Bangladesh, India and Pakistan. China has the largest population, but its government now manages to ensure that its people are reasonably fed. It has also recently managed to prevent any large increase in population.

Africa as a whole may not be overpopulated at present, but population density is putting pressure on land distribution in certain areas. In Kenya the population is increasing at about 3 percent per year. At this rate - among the highest in the world - the population will double in 25 years. The country may well have sufficient land, food-producing capacity and other resources to meet the demands of double or triple the present number of people. However, doubling food production is not enough. Kenya must also double the number of schools or school places, of hospitals or hospital beds, of houses and of all services in the 25 years that it will take for the population to double. Even then it will only have maintained the current level of development.

Each government must take its own decisions concerning population policy, but all governments must be aware that, if the nutritional status of people is to improve, the availability of food and services must increase more rapidly than the population.

Clearly when the number of people in a country, a community or a family increases, its food needs also increase. However, food availability is influenced by more shall population size. Economics, politics and geography are factors, too. Hong Kong and the Netherlands are both densely populated, yet they have little hunger and their infant and child mortality rates are low.

In most developing countries - even the poorest - in Africa, Asia and Latin America, infant and young child mortality rates have declined markedly in the past 30 years. When women continued to have the same number of babies and fewer died, family size increased.

In some countries, increased family size has also resulted from narrower spacing between pregnancies, younger age at first pregnancy and lack of knowledge about, or lack of availability of, family planning services. It is generally agreed that when the mother or the parents have confidence that most children born are likely to survive into adulthood, they are much more likely to consider and practice birth control.

Many of the more prosperous countries, particularly in Europe, have reached the stage of zero population growth, excluding growth from immigration. This means that the number of births per year nearly equals the number of deaths. In contrast, many developing countries have far more births than deaths and, consequently, rapidly increasing populations. However, several poor countries have reduced their rate of population increase, mainly through family planning methods.

Urbanization

Overall population growth is not the only demographic concern of many developing countries. The rapid increase in the percentage of people living in large cities is also a growing worry.

Population in urban areas has increased in part because of increased fertility rates, but migration from the rural areas to the cities is also a major cause. City dwellers in general are consumers, not producers, of food; as they become more numerous relative to the rural residents, the food production burden on the few becomes greater. In 1900 there were only four cities in the world with over 2 million residents; now there are over 100 such cities, as well as a number of megalopoli with over 10 million inhabitants.

The nutritional outcome of urbanization is on the whole positive. Urbanization, together with population growth and increasing incomes, contributes to tremendous increases in food demand and thus in the volume of food required, but also to varied and dynamic changes in dietary structure. The most significant dietary change caused by the urban migration has been the substitution of staple foods such as roots, tubers and coarse grains by other sources of energy such as highly milled cereals, sugar, soft drinks and other processed foods. In the urban environment, time constraints, availability of cheap, often subsidized processed foods and convenience of preparation are important considerations in influencing food consumption patterns.

The urban diet is generally more varied than the rural diet, mainly because of changes in non-staple foods. Fish, fresh vegetables, meat, poultry, milk and dairy products are consumed more often by urban people. Urban populations generally have lower energy intake than rural populations, but their physical activity may also be comparatively low. Consumption of animal protein, fat and vitamin A is higher in urban areas, and the iron consumed is better utilized. On the whole the diets of urban populations are more balanced than those of rural people.

A typical effect of urbanization is an increase in the amount of food eaten outside the home. Commercially prepared meals and other ready-to-eat foods are consumed from street vendors and food stalls. In many developing countries an informal sector for the sale of food has developed as a typically indigenous response to some of the food needs of the cities; this sector provides a cheap source of food and a significant source of income, particularly for women.

Urban nutrition is also affected by the fact that in most low-income urban households women work outside the home; as a consequence there has been an almost universal decline in breastfeeding in urban areas in all regions of the developing world, with a concomitant increase in the use of more costly breastmilk substitutes and commercial weaning foods.

On average, urban dwellers enjoy better nutritional status than their rural counterparts because of better health coverage and greater diversity in the diet. A more varied urban diet with minimal seasonal fluctuations confers important nutritional benefits. FAO data show that the incidence of child malnutrition, especially chronic malnutrition, is lower in urban areas. In Ghana, the weights of the adult urban population were found to be higher than those of the rural population. In general, urban areas also have lower morbidity and mortality rates, increased life expectancy, fewer children of low birth weight and fewer growth problems.

Technology

Despite rapid population growth, the world produced enough food in 1995 to feed adequately all the people on the globe - if the food were equitably distributed. Even if the population of the world doubles from the current 5 500 million to 11 000 million by the year 2030, the world's production will be capable of feeding all those people. Beyond that level, unless population growth stabilizes, serious shortages of food could result. Unlimited population increase on a planet of finite size is impossible; before too long the world would have standing room only, and each inhabitant one square metre of space.

It is a credit to agricultural advances and the skills of farmers that food supplies have increased to meet population needs. Many countries have achieved increased production levels not by expanding the land under cultivation, but by increasing the yields of cereals and other important crops per hectare farmed. This trend will have to continue. In addition, processing and marketing of foods must be improved.

Reproduction and Nutritional Status

In most developing countries the mean age of first menstruation is 12 to 24 months later than in industrialized countries. Menarche usually occurs about 12 months after the year with the greatest growth spurt (also known as peak height velocity). The onset of menses signals the beginning of a female's ability to become pregnant. It is almost certain that undernutrition delays the onset of menses. In this way poor nutrition influences human fertility.

Starvation and severe undernourishment, as in food shortages or famines resulting from drought, war or other factors, will usually result in cessation of menstruation in women of child-bearing age. Women who have ceased to menstruate in this way are infertile until their food intake improves. This is nature's way of preventing conception in undernourished individuals. Psychological consequences also result.

Numerous pregnancies and lactations, especially at short intervals, are likely to deplete the mother of nutrients unless she has an exceptionally good diet. Therefore, women with many children narrowly spaced are more likely to have poor nutritional status.

A woman whose diet is deficient during pregnancy, especially in terms of total food and energy, is likely to give birth to a baby that is smaller than it would have been if she were adequately nourished. Since mortality is more likely in underweight babies, a poor maternal diet is seen to increase the chances of death in the baby. Some studies, for example in Guatemala, have shown that when the diets of pregnant women were supplemented their infants had higher birth weights.

It has also been shown that a short interval between successive children may increase their risk of malnutrition and even of dying, particularly for fifth and subsequent children. Pregnancies that are too numerous and too narrowly spaced may be harmful to both the mother and the child. A mother practicing family planning simply to space her children more widely benefits also in nutrition and health.

Family planning is intimately related to health and nutritional status. Small family size, long intervals between pregnancies and gradual termination of breastfeeding are all associated with good health, positive nutritional status and even decreased mortality rates in the mother and family.

Breastfeeding, Fertility and Family Planning

For many years the idea that lactation prevented pregnancy was considered an old wives' tale. Now it is known as a scientific fact that women who are intensively breastfeeding their babies do experience a longer interval before menstruation begins again, and they are therefore less likely to have an early successive pregnancy than those who do not breastfeed. Breastfeeding is likely to lengthen birth intervals by an average of five to eight months. In this way the prolongation of full breastfeeding in developing countries is having a major effect in reducing fertility, in population control and in child spacing. Breastfeeding is nature's way of helping to space children. If bottle-feeding were to replace breastfeeding without the availability of contraceptives, the result would be major increases in the number of children born.

Contraceptive pills, especially high-oestrogen pills, may reduce a woman's ability to produce breastmilk. Therefore care must be taken in advising women to take contraceptive pills soon after childbirth. In contrast, it has been suggested that the intra-uterine device (IUD) may enhance or improve lactation.

Some contraceptives may have an effect on nutritional status. Certain contraceptive pills are believed to cause anaemia because they affect folate utilization. The IUD may cause increased bleeding, which can lead to iron deficiency anaemia.

It seems likely that a decrease in infant and child mortality is a prerequisite to the wide acceptance of family planning in those societies in which childhood deaths are common. Parents need to have confidence that their children will survive before they will risk limiting their family size. As malnutrition is one of the leading causes or contributory causes of death in children, it follows that improved nutrition will expedite the acceptance of family planning.

Improved nutrition is part and parcel of a better quality of life. Having fewer children in a family means more food, more room and less poverty; these also contribute to an improved quality of life. Wider spacing of children results in improvements in the health and nutritional status of children and their mothers. There is thus a circular effect.

There is much sense in linking nutrition and family planning activities, and even in integrating them into one programme. Both are related to maternal and child health and to total family health care. It may be advantageous for the same health personnel to deal with nutrition, family planning and maternal and child health. Some countries, such as Indonesia, where family planning is having a significant impact in reducing the rate of population increase and where families are smaller shall they were 20 years ago, have combined family planning activities with nutrition and health programmes; it appears to have worked well.

Nutritionists may be concerned about the alarming rate of increase of the world's population. Kenya, for example, had a population of 26 million in 1994 and will have over 50 million people in 2020; nutritionists may be alarmed by all the implications of this population growth in terms of land shortage and mushrooming urban slums. In their work, however, nutritionists usually deal with problems of families or communities. It is important then to help people, especially women and their partners, understand the benefits of smaller families and the fact that more children require more resources: more food, more care, more time, more school fees, more money and so on. The appropriate strategy may be to persuade them that children today have a better chance of surviving than children had in 1955, and that quality of life is more important than numbers of children.

It is of particular importance in many developing countries first to empower women to control their own fertility and to be in a position to have the numbers of children they desire, and second to influence men to respect these rights of their female partners. The onus of having more children falls on the whole family, but in most countries by far the greatest additional burden of work falls on the mother. It is she who must endure pregnancy for another nine months, breastfeed the infant, drain her own health and perhaps impair her own nutritional status.

The education of girls and the empowerment of women to earn money, control resources and have more independence are all achievements that generally lead women to control their own fertility and have fewer babies. Women's support groups, sex education in schools, involvement of men in discussions, later marriage and more intensive breastfeeding are all likely to reduce the mean number of babies produced per mother.

Nutrition workers, be they in the field of health, agriculture, education or social services, should make themselves conversant with modern methods of family planning. They should be able to discuss these methods with people either individually or in groups, and they should know how to advise people to use local family planning services. If these services are inadequate or cause problems for women or families, nutrition workers should be advocates for improved family planning services. The more the choices available to women and men, the more likely it is that the babies born will be wanted babies, Of course workers must respect national laws and cultural norms. If abortion is illegal, the law will need to be respected. Communities have been most successful in limiting unwanted births where several family planning methods are available: contraceptive pills, condoms, lUDs, male and female surgical sterilization and, if legal, well-conducted abortions. In some countries newer methods such as hormone-releasing implants (Norplant) or the so-called abortion pill may be expected to help in family planning in the next five years.

Nutrition During Particular Times in the Life Cycle

Nutrient requirements differ to some extent at different periods in the life cycle. Females of reproductive age have extra needs because of menstruation and, of course, during pregnancy and lactation. Infants and children have greater requirements on a unit weight basis than adults, mainly because they are growing. Older people are also a vulnerable group; they are at greater risk of malnutrition than younger adults.

Humans get energy from the foods and liquids they consume. The nutrient requirements of women of reproductive age (especially during pregnancy and lactation), of young children and adolescents and of older people are different from those of men between the ages of 15 and 60; therefore all people do not need the same amounts of food.

Women of Reproductive Age

Women of child-bearing age have certain nutritional needs above those of adult males. One reason is that the loss of blood during menstruation leads to a regular loss of iron and other nutrients and makes women more prone than men to anaemia. In addition, however, in many developing countries women work much harder than men. In rural areas they are often heavily involved in agriculture, and in urban areas they may work long hours in factories and elsewhere; yet when they return home from the field or the factory they still have much work to do in the household, including food preparation and child care. Frequently the heavy burden of collecting water and fuel falls on women. All of this labour increases women's needs for nutritional energy and other nutrients.

The nutritional status of women before, during and after pregnancy contributes a good deal to their own general well-being, but also to that of their children and other members of the family. The field of maternal nutrition focuses attention on females as mothers. It has often concentrated on their nutritional status mainly as it is related to the well-being of the infants that they produce and their ability to breastfeed, nurture and raise their children. The health and well-being of the mother herself has been relatively neglected. Similarly, the field of maternal and child health has put major emphasis on the child and on providing services and help to women mainly so that they can have successful pregnancies and lactations; this is also in the interests of the infant, without much concern for the mother. The dual role of women as mothers and productive workers is compromised by poor diets and ill health; not only their own well-being but that of the whole family is affected. A heavy work load may push a woman with marginal food intake over the brink and into a state of malnutrition.

A poor diet, frequent acute and some chronic infections, repeated pregnancies, prolonged lactation and a heavy burden of work may all contribute to serious physiological depletion and sometimes to overt malnutrition. The term "maternal depletion syndrome'' has been suggested. In many countries young women in their late teens appear hearty, happy, healthy and attractive, but only ten or 15 years later, as young women in their late thirties, they are prematurely old, tired, down-trodden and unhealthy. Too often, the young female does not even live out her teens before her first pregnancy.

Pregnant Women

During pregnancy a woman's nutritional needs become greater shall at other times in her life. Her diet needs to provide all the elements needed for the growth of a fertilized ovum or egg into a viable foetus and baby. As the woman nourishes herself she also nourishes the growing foetus as well as the placenta to which the foetus in her uterus is attached by its umbilical cord. At the same time her breast tissue prepares for lactation.

During the first half of pregnancy extra food is needed for the mother's uterus, breasts and blood - all of which increase in size or amount - as well as for the growth of the placenta. The increased need for food continues in the last half of the pregnancy, but during the last trimester the extra nutrients are required mainly for the rapidly growing foetus, which also needs to develop nutrient stores, particularly of vitamin A, iron and other micronutrients, and energy stores of fat. An adequate diet during pregnancy assists the mother to gain the extra weight that is physiologically desirable and helps ensure that the baby's birth weight is normal.

Healthy women gain weight during pregnancy if they are not overworked. Just as a heavy person needs more energy to perform the same amount of physical work as a lighter person, a pregnant woman also needs more energy. In industrialized countries many women have an easy life during pregnancy; they rest frequently, thus reducing their energy needs. However, in much of Africa and some other regions, pregnant women remain active, even during their last few months of pregnancy. The Basal metabolic rate (BMR) usually increases during pregnancy, which also raises energy requirements. Thus most women need more energy when they are pregnant, even if they are not overworked. For the overburdened woman of the developing world, who gets little rest and not much food, weight loss is a real and dangerous prospect.

There is little doubt that abortions, miscarriages and stillbirths are more common in women who are poorly nourished than in those who are adequately nourished. Dietary deficiencies probably also increase the risk of producing a malformed foetus. Severe malnutrition reduces fertility and therefore the likelihood of conception. A severely malnourished woman ceases to menstruate. This is clearly a natural device to stop the loss of nutrients in the menstrual flow and to protect the woman from the rigours of pregnancy and childbirth. Nevertheless, there is little evidence of lack of fertility among the less severely malnourished, and mildly malnourished women are the majority in Asia and parts of Africa.

The weight of the infant at birth is influenced by maternal nutrition. Low birth weights can be expected of infants born to malnourished mothers. Even a modest increase in energy intake during pregnancy tends to increase the birth weight of the infant.

In many developing countries 50 to 75 percent of pregnant women have anaemia. Anaemia often contributes to high maternal mortality rates. All pregnant women should attend a clinic at regular intervals for antenatal examination which should include checking of haemoglobin levels. Practical advice should be given regarding diet, taking into account what foods are locally available and what the mother can afford. It is accepted policy in many countries that pregnant women be advised to take medicinal supplements of iron, or sometimes iron-folate.

In areas where vitamin A deficiency is known to be a public health problem, infants of mothers who have poor vitamin A status are born with low vitamin A stores.

A diet with adequate amounts of vitamin A is clearly important during pregnancy, both for the mother and the baby. However, high medicinal doses of vitamin A, as are given to young children, are not recommended during pregnancy. In the case of many other nutrients, however, the child is truly parasitic and takes all the nutrients it requires irrespective of whether the mother has a deficiency or not.

In some cultures there is a fear that extra food given during pregnancy will make the baby too large and thus cause a more difficult or complicated delivery. This is not true for healthy women of normal size. Women of short stature or with a contracted pelvis may have difficulty in delivering babies and may require special care before and during delivery.

At the time of birth the mother loses blood, not infrequently 500 to 1000 ml, and she needs nutrients to regenerate that blood.

Lactating Mothers

In most developing countries the majority of women breastfeed their newborn infants for a period of weeks or months after delivery. The nutritional stores of a lactating woman may already be more or less depleted as a result of the pregnancy and the loss of blood during childbirth. Lactation raises nutrient needs, mainly because of the loss of nutrients first through colostrum and then through breastmilk.

Breastmilk volume varies widely, but for fully breastfed babies around four months of age, it is often 700 to 800 ml per day. It may rise later to as much as 1 000 ml or more. The nutrients present in this milk come from the diet of the mother or from her nutrient reserves. It is recommended that mothers exclusively breastfeed their infants for six months and then begin to introduce other food while continuing to breastfeed for as long as they wish, often into the second year or beyond.

During the period of exclusive or full breastfeeding the woman usually will not menstruate. The duration of amenorrhoea varies from as little as four months to as long as 18 months or more. During that time the lactating woman will not be losing the iron normally lost with each menstrual period.

The conversion of nutrients in food to nutrients in breastmilk is not entire. In the case of energy, it is about 80 percent, so for every 800 kcal in breastmilk the mother needs to consume 1 000 kcal in her food. To have good nutritional status the breastfeeding woman has to raise nutrient intake.

There is a widely held belief that the composition of breastmilk varies enormously. This is not so. Human breastmilk has a fairly constant composition, and is only selectively affected by the diet of the mother. One litre of milk provides about 750 calories and contains approximately the following:

— 70 g carbohydrate,
— 46 g fat,
— 13 g protein,
— 300 mg calcium,
— 2 mg iron,
— 480 μg vitamin A,
— 0.2 mg thiamine,
— 0.4 mg riboflavin,
— 2 mg niacin,
— 40 mg vitamin C.

The fat content of breastmilk varies somewhat. The carbohydrate, protein, fat, calcium and iron contents do not change much even if the mother is short of these in her diet. A mother whose diet is deficient in thiamine and vitamins A and C, however, produces less of these in her milk. Thiamine deficiency in the lactating mother can lead to infantile beriberi in the baby. In general the effect of very poor nutrition on a lactating woman is to reduce the quantity rather than the quality of breastmilk.

Lactating mothers should be encouraged to attend a clinic with their babies during the months after delivery. At the clinic both mother and baby should be examined. The mother should have her haemoglobin level checked and also her weight. Medicinal iron in the same quantities as recommended during pregnancy should be given. The mother should be given advice on consuming a mixed diet. This is also a good time to discuss the mother's desire for further pregnancies and her view of the ideal spacing between pregnancies and to provide information and help regarding family planning. Relatively wide spacing between births is usually to the nutritional advantage of the mother the infant and even the next foetus. Narrow spacing between births prevents the mother from restoring her nutrient reserves before the next pregnancy, provides her with more work and a shorter time to care for her infant exclusively, and may influence her to breastfeed for a shorter period than is desirable.

At each postnatal visit both the mother and the baby should be examined, and advice on the diets of both mother and infant should be provided. A satisfactory gain in the infant's weight is the best way to judge the adequacy of the diet of the infant. In the first few months when there is exclusive breastfeeding the infant's adequate weight gain is a clear indication that the mother is producing sufficient breastmilk. Almost all mothers can successfully breastfeed their infants.

Infants and Preschool-age Children

Provided that the mother has adequate breastmilk, breastfeeding alone with no added food or medicinal supplementation is all that is needed for the normal infant during the first six months of life. Exclusive breastfeeding means that not even water, juice or other fluids are provided; none of these are needed. The infant should be examined regularly at the clinic, where weight gain is seen to indicate adequate nutrition. At the clinic a schedule for immunization will be set up, and this needs to be followed. Infants born with low weight (because of prematurity, for example) or twins may need special attention, and possibly iron or other supplements should be given. Up to six months of age many breastfed infants have considerable natural immunity to many infections.

As the children get older they gain weight and length. The increased energy requirements are based more on the weight of the child than on the age. Because healthy, well-nourished children follow a growth pattern, however, there is a close correlation between recommendations based on age and those based on weight. A baby 2.5 months of age weighing 5 kg requires 5 x 120 kcal = 600 kcal, whereas a baby eight months of age weighing 8 kg requires 8 x 110 kcal = 880 kcal.

At six months of age complementary feeding should be introduced gradually while the infant continues to be breastfed intensively and to receive most of his or her energy and other nutrients from breastmilk and not from complementary foods. From six to 12 months, it is highly desirable that breastfeeding should continue and that the child should get as much milk as possible from the mother while other foods, first semi-solid and then solid, should be introduced to the diet of the infant for normal growth and health.

Breastmilk is relatively deficient in iron, and the infant's store of iron is sufficient only until about six months of age. From six to 12 months, the normal infant may be expected to gain between 2 and 3 kg. The infant, while continuing to receive breastmilk, will now need foods to provide extra energy, protein, iron, vitamin C and other nutrients for growth.

The needed energy can usually be obtained from a gruel of whatever is the local staple food. The quantity and bulk can profitably be reduced if some edible oil or fat-containing food is also eaten. If the staple is a cereal such as maize, wheat, millet or rice, it will also provide a useful quantity of protein, but if it is plantain or a root such as cassava or yam, it will supply very little protein. In this case, once relatively little breastmilk is being consumed it is important to provide extra protein-rich foods from those available to the family.

In the 1950s and 1960s it was thought to be very important that complementary foods and then foods given after termination of breastfeeding should include animal protein in large amounts. This has been shown to be unnecessary. In developing countries these foods are often too expensive for poor families or are unavailable. More important is the need to feed the young child frequently, with foods that are not too bulky and are both nutrititious and of high energy density.

Legumes such as beans, peas, lentils, cowpeas and groundnuts are good sources of protein and should be added to the diet of the child. They can be ground or crushed before or after cooking.

The above foods, as well as providing energy and protein, will also provide some iron. Additional iron can be obtained from edible green leaves, which also contain carotene and vitamin C. Carotene and vitamin C can also be obtained from fruit. Ripe papayas and mangoes are excellent sources and are usually most acceptable to young children. Vitamin C can alternatively be provided by citrus fruits (e.g. Oranges) or other fruits (e.g. guavas).

Gradually, as more teeth erupt, the child can be put on a more solid diet. By the age of two years, the child may have stopped breastfeeding and may be completely weaned.

The term "weaning" has been used to describe the introduction of foods and fluids other than breastmilk and the transition to a solid diet without breastmilk. However, people in Northern countries also talk of "weaning from the bottle". The word is therefore often misunderstood, and it may be better not to use it because of the confusion it causes. Rather, the transition can be described as four stages:

— the first four to six months when all the infant's nutrients come from breastmilk;

— the next few months when just as much (or more) breastmilk is provided but other appropriate, often soft, nutritious foods are introduced in increasing amounts, with efforts to prevent these from causing a decline in breastmilk consumption;

— the next stage, perhaps starting at about 12 to 15 months, when the baby is still breastfeeding but is getting considerably more of his or her nutrients from nutritious foods - most of them ordinary village or family foods than from breastmilk;

— the end of breastfeeding, the stage termed "sevrage" (a good French term literally meaning "severance from the breast"), which can occur as late as the mother wants, sometimes when the infant is over two years of age.

After sevrage appropriate family foods are provided. These need to be nutritious, suitable for the child, energy dense and given frequently, perhaps four to six times per day, not just in two or three meals per day as may be the family practice. The young child should be fed between family mealtimes if these are limited to two or three per day.

The mother responsible for feeding a toddler who is no longer breastfeeding must keep in mind that the child, whether boy or girl, has special needs.

The proper feeding of a toddler requires time and patience. Special utensils or equipment are not necessary, but a sieve or strainer is useful. Adult foods can be chopped up and forced through a strainer into a cup or on to a plateful of gruel for the child. A strainer can readily be made if none is available. Otherwise, various foods can be crushed before cooking using a pestle and mortar, which are found in most households.

In some societies gruel or porridge made from the local staple is made sour or partially fermented. This is a good practice. Small amounts of germinated cereal seeds, often millet or sorghum, are crushed and added to maize or other porridge. The amylase present breaks down some of the starch, causing the porridge to become thinner (more liquid), so it is easier for the young child to consume, and making it more energy dense. The food is also safer, because the growth of disease-causing organisms is inhibited in sour or fermented gruel. Some societies sour children's foods by addition of lime or lemon juice. This also is advantageous, and enhances the absorption of iron.

The period from six to 36 months of age is of paramount importance nutritionally. The mother should take the child regularly to a clinic if one is available. The happiness, general appearance and weight of the child are the best general indicators of adequate nutrition. Many children of this age in developing countries do not grow at the rate they should, and some develop protein-energy malnutrition.

The first three years of life are also those when the important micronutrient deficiencies of vitamin A and iron are most likely to occur in children. From three years of age the risks are reduced, but in many parts of the world growth continues to lag, incidence of intestinal worms and other parasitic diseases may increase and other nutrition and health risks arise.

From three years of age onwards the child has usually stopped breastfeeding and is consuming family foods. The child can now obtain adequate nutrients in three meals per day, but until the child reaches the age of five years, parents should make certain that the child is eating adequately and getting his or her fair share of the most desirable foods, which may well be those that are most tasty and in shortest supply. Special attention may need to be given when children have a poor appetite or when they are ill and their appetite is reduced. For the whole family, but especially for children, care must be taken that food, water and other fluids are safe and not contaminated. Good personal and household hygiene are of the greatest importance. Washing hands with soap and water before meals or food handling is a good family rule.

Parents should understand the needs of the child and see that the right foods are available in adequate quantities and prepared in palatable ways

As children increase in weight and age they need more food to provide them with more energy and more of the other nutrients essential for growth and health. Thus a child aged six to 12 months and weighing 8.5 kg requires 950 kcal per day, whereas a child aged five to seven years weighing 19 kg requires 1 820 kcal [almost twice as much) and a boy aged 17 years weighing about 60 kg requires 2 770 kcal (almost three times as much).

Mothers need to understand that as children grow beyond infancy, they increase in weight and require more food to eat.

School-age Children

The vast majority of schoolchildren in developing countries attend primary schools. Most are at day schools, few of which provide a midday meal. In rural areas the school is often some kilometres from the parents' home. The child frequently has to leave home early and walk a considerable distance to school. Often the child has little or no breakfast at home before he or she sets out; there is no meal at school; and the first, and sometimes only, meal of the day is late in the afternoon.

The nutritional needs of a schoolchild are high. The adolescent child has proportionately higher requirements for most nutrients than the average adult. It is practically impossible for an adolescent to obtain adequate quantities of the right foods from one or even two meals a day. It is highly desirable that school-age children cat some food before going to school and some food at school, or during the middle of the day outside the school grounds, as well as the food eaten at home.

Food before Going to School

It is not practical for many mothers to rise before dawn to spend the considerable time necessary to light a fire and prepare a hot meal for children before school-time. Therefore, if no hot breakfast is available, some fruit, cold cooked potatoes, rice, cassava or even cold porridge should be left over from the previous day for the schoolchild to eat before leaving home in the morning. In some areas cold chapattis, tortillas or wheat products such as bread may be available.

Food Eaten at School

This may consist of a midday school meal or a snack taken to school.

A midday school meal is the ideal. It should provide reasonable amounts of the nutrients most likely to be missing or short in the home diet. A whole-grain cereal as the basis and a side dish of legumes with vegetables or green leaves make an excellent school meal. There are many possibilities, depending on what foods are locally available. The meal might include some protein-rich food and some food containing vitamins A and C.

School meals are beneficial because they often supply much-needed nutrients; they can form the basis for nutrition education; they are a good way of introducing new foods; and they prevent hunger and malnutrition. School meals, in addition to improving nutritional status, may increase enrolment, especially for girls, and may reduce absenteeism. However, in many developing countries, for many reasons, school meals are unavailable. Parents' organizations can sometimes work with teachers to organize community school feeding or food supplementation or nutritious snacks. School meals can provide a good environment for nutrition education. Further nutrition education can be carried out as an extracurricular project. A school vegetable garden or orchard can provide foods with valuable extra nutrients for the midday meal. Poultry keeping, small animal production (rabbits, guinea-pigs, pigeons, etc.) and fish pond construction, in areas where they are suitable, are educative projects and can provide food for a school meal

A midday school meal might be provided by the government or local authority as part of the education system and could be paid for from the normal school fees Alternatively, a midday meal system might be started and paid for from special fees collected from the pupils daily, weekly or per term. Local organizations might provide certain food items free or at low prices for school feeding, thus reducing the overall cost.

The cost of school feeding can be reduced by local self-help efforts on the part of villagers, parents' committees and pupils These efforts may fit in well with self-help community projects For example, a small kitchen shelter can be built on a self-help basis Instead of a paid cook, a rota of parents can take turns doing the cooking. Pupils can collect fuelwood at weekends However, it must be stressed that the provision of a midday school meal must not detract from the parents' responsibilities to provide a good diet for schoolchildren at home.

In the absence of a school lunch, parents should send their children to school with some food to be eaten at midday. However, they may have real difficulty in finding suitable foods The various foods suggested for a cold breakfast can equally provide the solution for a midday snack The sort of food taken will vary according to what is available locally Possibilities include a few bananas, cooked whole cassava, sweet or ordinary potatoes roasted in their skins, fruit, tomatoes, roasted maize on the cob, roasted groundnuts, coconuts, cold grilled fish, smoked cooked meat, hard-boiled eggs, a calabash of sour milk or some bread, a chapatti or tortillas.

Some schools above primary level are boarding schools These usually provide three meals a day, and the menu should be based on recommendations made to the school by someone with dietetics training Occasionally schools plead lack of money as an excuse for an inadequate diet School meals need not be luxurious, but they should be balanced and should provide all the nutrients necessary for growth and health The child with an inadequate diet will not only fail to grow properly, but may also develop anaemia and other signs of malnutrition and will not be able to concentrate on or benefit fully from the education provided.

Increasingly in urban areas, and even to some extent in more heavily populated rural districts, entrepreneurs set up stalls and the like near schools so they can prepare and sell foods to schoolchildren. These "street foods" often have the advantage of providing access to cooked foods at relatively low cost, but the disadvantages include poor hygiene, poor-quality food and high prices. Where the main source of a midday snack or meal for primary or secondary schoolchildren is a vendor, the food is available only to children who have money to purchase it. Often the wealthier children participate and the children from the poorest families, or those whose parents will not provide money, do not.

The health of schoolchildren also needs consideration. In many countries school health services are non-existent or very poor. Examination for sight and hearing defects is important. Routine deworming might be initiated. Attention to micronutrient deficiencies may be needed in areas where children are at risk of iron, vitamin A or iodine deficiency. Iodine is especially important when girls reach puberty and before they have their first pregnancy.

Unfortunately, in some countries a large percentage of school-age children do not attend school. In some countries far more boys than girls attend school. Out-of-school children have the same nutritional and health needs as children attending school, but they do not benefit from school meals and other services. They are an often forgotten and relatively neglected group of the population, including children from the poorest families as well as children with disabilities, either physical or psychological.

Older Persons

Older people, like all others, need a good diet that provides for all their nutrient needs. In more affluent societies, older adults are often plagued with chronic diseases that have nutritional origins or associations. These conditions include, among others, arteriosclerotic heart disease, sometimes leading to coronary thrombosis; hypertension, which may lead to stroke or other manifestations; diabetes, with its serious complications; osteoporosis, which frequently leads to hip fracture or collapse of vertebrae; and loss of teeth because of dental caries and periodontal disease. These diseases are rapidly becoming more prevalent in developing countries.

Many older people, especially if unfit, take less exercise and so may need less energy.They may, therefore, eat less food and as a result get fewer micronutrients, but their needs for micronutrients are unchanged. Consequently, conditions such as anaemia are common. Older people who have lost many or all of their teeth or who have gingivitis or other gum problems may find it difficult to chew many ordinary foods and may need softer foods. Fed on a normal family diet, they may eat too little and become malnourished. They may also suffer from illnesses which reduce their appetite or desire for food, which may also lead to malnutrition.

In many rural traditional societies old people are cared for at home by relatives and others in the community. By contrast, many older people in the richer, industrialized countries of the North live lonely lives and are relegated to old people's nursing homes and other unpleasant institutions. In some developing countries the traditional support systems and extended families are breaking down, especially with urbanization and migration, and old people there may end up lonely, living in poverty, with chronic illnesses, poor hearing and vision and perhaps psychological problems. Compounding these problems, they will face difficulties in producing food, purchasing it and preparing it.

Many of the older people are poor women, who are especially vulnerable. They are members of society in special need of both good care and a good diet, just as children are in their early years.

In some countries special services are established to help older or poor people obtain food in soup kitchens or in their homes. These services can be helpful. Preferable, however, would be community and family efforts to care for older people who cannot care for themselves and who are at risk of malnutrition and disease.

BREASTFEEDING

For most of human history nearly all mothers have fed their infants in the normal, natural, no-fuss way: breastfeeding. Most traditional societies in Africa, Asia and Latin America have had good local knowledge about breastfeeding, although practices have varied from culture to culture.

The famous paediatrician Paul Gyorgy said, "Cows' milk is best for baby cows and human breastmilk is best for human babies". No one can deny the truth of that statement. It is increasingly acknowledged, therefore, that every mother has the right to breastfeed her baby and every infant has the right to be breastfed. Any obstacles placed in the way of breastfeeding are an infringement of these rights; yet in almost all countries there are many babies who are not breastfed or are breastfed for a relatively short time.

In recent years interest in breastfeeding has grown. Part of the reason is the much-publicized controversy over the replacement of breastfeeding by bottle-feeding and the related aggressive promotion of manufactured breastmilk substitutes by multinational corporations. The womanly art of breastfeeding has in recent years been rediscovered in Europe and to a lesser extent in North America. Unfortunately, however, use of bottle-feeding continues to increase in many non-industrialized countries of the South. The most serious consequences of this shift from breast to bottle are seen among poor families in Africa, Asia and Latin America.

Advantages of Breastfeeding

Extensive studies comparing the composition and relative benefits of human milk and its substitutes have been published over the past 50 years and especially in the last decade. Most of the new research has supported the many advantages of breastfeeding over other methods of infant feeding. A vast body of research from all over the world sustains the recommendation that only breastmilk be fed to infants for the first six months of life. Certainly in developing countries where the risks of complementary feeding usually outweigh any possible advantages, breastfeeding alone up to six months of age is advised.

The advantages of breastfeeding over bottle-feeding and the reasons wily it is so strongly recommended are summarized as follows.

— Breastfeeding is convenient; the food is readily available for the infant, and no special preparation or equipment is needed.
— Breastmilk provides a proper balance and quantity of nutrients ideal for the human infant.
— Both colostrum and breastmilk have anti-infective constituents that help limit infectious.
— Bottle-feeding enhances the risk of infections from contamination with pathogenic organisms in the milk, the formula and the water used in preparation, as well as in bottles, teats and other items used for infant feeding.
— Breastfeeding is more economical shall bottle-feeding, which involves costs for infant formula or cows' milk, the bottles and teats and the fuel necessary for sterilization.
— Breastfeeding prolongs the duration of post-partum anovulation, helping mothers to space their children.
— Breastfeeding fosters enhanced bonding and relationship between mother and infant.
— An apparent lowered risk of allergies, obesity and certain other health problems is seen in breastfed infants compared with those who are artificially fed.

There is now overwhelming evidence of the health advantages of breastfeeding as indicated by lower infant morbidity and mortality than for bottle-fed infants. The advantages accrue mainly to tile two-thirds of the world's population who live in poverty, although some studies have shown lower rates of diarrhoea and other infections and less hospitalization among breastfed infants even in affluent communities. There is now evidence that women who breastfeed their infants have a reduced risk of breast cancer, and perhaps of uterine cancer, than women who do not.

Problems with Bottle-feeding

An infant who is not breastfed, or even one who is not exclusively breastfed for the first four to six months of life, loses many or all of the advantages of breastfeeding mentioned above. The most common alternative to breastfeeding is bottle-feeding, usually with a manufactured infant formula, but not infrequently with cows' milk or other liquids. Less commonly an infant in the first four to six months of life is fed solid foods in place of breastmilk. Some mothers do use a cup and spoon rather than a bottle to provide cows' milk, infant formula or gruel to young babies. Spoon-feeding has some advantages over bottle-feeding but is much less satisfactory than breastfeeding.

Infection

Whereas breastmilk is protective, alternative infant feeding methods increase the risk of infection, mainly because contamination leads to the increased intake of pathogenic organisms. Poor hygiene, particularly with bottle-feeding, is a major cause of childhood gastro-enteritis and diarrhoea Infant formula and cows' milk are good vehicles and culture media for pathogenic organisms. It is incredibly difficult to provide a clean, let alone sterile, feed to an infant from the bottle under the following circumstances:

— when the family water supply is a ditch or a well contaminated with human excrement (relatively few households in developing countries have their own safe supply of running water);
— when household hygiene is poor and the home environment is contaminated by flies and faeces;
— when there is no refrigerator or other safe storage space for reconstituted formula or for cows' milk;
— when there is no sum-on stove and on each occasion someone has to gather fuel and light a fire to boil water to sterilize a bottle;
— when there is no suitable equipment for cleaning the bottle between feeds and when the bottle used may be of cracked plastic or an almost uncleanable soda bottle;
— when the mother is relatively uneducated and has little or no knowledge of the rode of germs in disease.

Malnutrition

Artificial feeding may contribute importantly in two ways to protein-energy malnutrition (PEM) including nutritional marasmus. First, as discussed earlier, formula-fed infants are more likely to get infectious including diarrhoea which then contribute to poor growth and PEM in infancy and early childhood. Second, mothers in poor families often overdilute infant formula. Because of the high cost of breastmilk substitutes, the family purchases too little and tries to stretch it by using less than the recommended amounts of powdered formula per feed. The infant may be given the correct number of feedings and the recommended volume of liquid, but if it is too dilute each feed may be too low in energy and other nutrients to sustain optimal growth. The result is first growth faltering and then perhaps the slow development of nutritional marasmus.

Economic problems

A very important disadvantage of formula feeding is the cost for the family and for the nation. Breastmilk is produced in all countries, but infant formula is not. infant formula is a very expensive food, and if countries import it, then foreign exchange is unnecessarily spent. Choosing breastfeeding over bottle-feeding, therefore confers significant economic advantages for families and for poor countries.

Infant formula is a better product for a one-month-old baby shall fresh cows' milk or whole milk powder. Dried skimmed milk (DSM) and sweetened condensed milk are contraindicated. However, infant formula is extremely expensive relative to the incomes of poor families in developing countries. In India, Indonesia and Kenya it would cost a family 70 percent or more of the average labourer's wage to purchase adequate quantities of infant formula for a four-month-old baby. The purchase of formula as a substitute for breastmilk diverts scarce family monetary resources and increases poverty.

For many countries that do not manufacture infant formula a decline in breastfeeding means an increase in tile importation of manufactured breastmilk substitutes and the paraphernalia needed for bottle-feeding, These imports may lead to a worsening of the already horrendous foreign debt problems for many developing countries. Even where infant formula is locally made, the manufacture is frequently controlled by a multinational corporation, and profits are exported. Therefore, the preservation of breastfeeding or a reduction in artificial feeding is in the economic interests of most developing countries. Economists and politicians may be more inclined to support programmes to promote breastfeeding when they appreciate that such measures will save foreign exchange; the economic implications are often of more interest to them than arguments about the health advantages of breastfeeding.

Properties and value of Breastmilk

Immediately after giving birth to a baby, a mother produces colostrum from both breasts. Within a few days the milk "comes in", and it increases in quantity to match the needs of the baby. A mother's milk production is influenced mainly by the demands of her baby, whose sucking stimulates breastmilk secretion. The more the baby sucks, the more milk the mother will produce. The amount will often increase from about 100 to 200 ml on the third day after the baby's birth to 400 to 500 ml by the time the baby is ten days old.

Production can continue to increase to as much as 1 000 to 1 200 ml per day. A healthy, normally growing four-month-old infant of average weight will, if exclusively breastfed, receive 700 to 850 ml of breastmilk in 24 hours. Provided the babies can suckle as much as they want, they will always get enough milk. This is probably the only time in life when a person can eat as much of what he or she likes whenever he or she likes! Feeding on demand - any time, day or night - is the traditionally practiced method of breastfeeding. It is best achieved by a mother who is happy, relaxed, confident and free to be with her baby all the time. In these circumstances, the mother and baby form what has been termed a dyed - a special twosome.

One litre of breastmilk produces about 750 kcal. Cows' milk provides about three times more protein and four times more calcium, but only about 60 percent of the carbohydrate present in human breastmilk.

Most studies now clearly indicate that the nutrients present in milk from a healthy, well-nourished mother satisfy all the nutritional needs of the infant if the infant is consuming enough milk. Even though the iron content of breastmilk is low, it is sufficient and well enough absorbed to prevent anaemia during the first four to six months of life. Cows' milk is even lower in its iron content and is not very well absorbed by the baby, so infants fed cows' milk are very likely to develop iron deficiency anaemia.

Breastmilk may vary somewhat between individuals, and probably to some minor extent in different parts of the world. It is also different at the beginning and end of each feed. The so-called foremilk is more watery and contains less fat in comparison with the milk of the latter part of the feed, which is somewhat thicker and whiter in appearance and more energy dense because it contains more fat.

Of particular importance is the presence in colostrum and breastmilk of anti-infective factors (which are not present in infant formula). These include:

— antibodies and immunoglobulins, some of which work in the baby's intestines and prevent disease-causing organisms from infecting the baby;

— living cells, mainly white blood cells, which may produce important substances such as interferon (which may fight viruses), immunoglobulin A, lactoferrin and Iysosomes;

— other factors such as the bifidus factor which helps certain friendly bacteria such as lactobacilli to grow and proliferate in the infant's intestines, where they help ensure an acid environment (from lactic acid) which discourages the growth of harmful organisms.

In simple terms, breastmilk leads to an environment in the intestines of the baby that is harmful and unfriendly to disease-causing organisms. The stool of a breastfed infant differs in appearance from that of a formula-fed baby.

Science and industry have combined to produce breastmilk substitutes which are intended to mimic breastmilk in terms of quantities of known nutrients present in mothers' milk. These products, often called infant formulas, are the best alternative to breastmilk for those few babies who cannot receive breastmilk. All infant formulas are based on mammalian milk, usually cows' milk. Even though infant formulas may be the best alternative to human breastmilk, they are not the same. They do include the known nutrients that are needed by the infant, but they may not include those nutrients that have not yet been identified; in this case it is not possible to know what the bottle-fed infant is missing. Indeed, in some respects infant formulas are so different from human milk as to be at best unsuitable and at worst dangerous. The manufactured milks do not have the anti-infective properties and living cells that are present in human milk. The manufactured products may cause the infant to suffer health problems that would never be brought on by human milk.

Breastmilk, particularly because of the immunoglobulins it contains, seems to protect babies against allergies. In contrast, the non-human and cow proteins present in breastmilk substitutes, as well as other substances which enter infant formulas during manufacture, may provoke allergies. The important end result is a much higher rate of eczema, other allergies, colic and sudden infant death syndrome (SIDS) in formula-fed infants than in breastfed infants.

Colostrum

Colostrum is the yellowish or straw-coloured fluid produced by the breasts for the first few days after the birth of the baby. Colostrum is highly nutritious and rich in anti-infective properties. It could be said that the living cells, immunoglobulins and antibodies in colostrum constitute the infant's first immunization.

In most societies, colostrum is recognized to differ from breastmilk because of its colour and its creamy consistency, but its enormous value to the baby is not universally acknowledged. In many parts of the world mothers do not feed colostrum to their babies; they wait until white milk is secreted from the breast. Some mothers (and grandmothers) think that in the first days after birth the newborn infant should receive other fluids or foods, for example, tea in India, jamus (traditional medicinal potions) in Indonesia and sugar or glucose water in many Western hospitals. These foods are not needed and are in fact contraindicated. The baby at birth has adequate body water and fluids and enough nutrients, so the only feeding needed is colostrum and then breastmilk for the first four to six months of life.

Infant Feeding Trends

The percentage of mothers who breastfeed their infants and the duration of breastfeeding varies among countries and within them. Exclusive or near exclusive breastfeeding for the first four to six months of life, followed by breastfeeding for many more months while other foods are introduced, is considered by scientists to provide optimum infant feeding. This ideal, however, does not exist in any country, North or South.

Most mothers in traditional societies, particularly in rural areas in developing countries, still breastfeed all their children for a long time. Few, however, practice exclusive breastfeeding, and many do not provide colostrum to their babies.

Many mothers in Europe and North America, by contrast, do not breastfeed their children. The trend away from breastfeeding was most marked in the 1950s and 1960s, when fewer than 15 percent of American babies two months of age were breastfed. During those years a marked decline in breastfeeding was reported from some Asian and Latin American countries. By the mid-1990s there was a modest resurgence of breastfeeding in the industrialized countries of the North, especially among, better-educated mothers. In poor Asian, African and Latin American countries breastfeeding rates are often lower in urban areas and higher in the rural areas where people have less education.

There are many reasons for a decline in breastfeeding or for the unnecessary use of breastmilk substitutes, and the reasons vary from country to country. Aggressive promotion by the manufacturers of breast-milk substitutes is one cause. Promotional practices have now been regulated in many countries, but the manufacturers continue to circumvent the accepted codes of conduct and to promote their products, even though such practices may contribute to infant morbidity.

Actions by the medical profession have also contributed to the reduction in breastfeeding. In general, health care systems in most countries have not adequately supported breastfeeding. Even in many developing countries doctors and other health care professionals have had a negative role and have contributed to reduced levels of breastfeeding. This situation is changing, but many health professionals are still relatively ignorant about breastfeeding

Breastfeeding often declines when rural women move to urban areas, where traditional practices may get replaced by modern ones or be influenced by urbanization. Women who take jobs in factories and offices may come to believe that they cannot combine their employment with breastfeeding, and labour conditions and labour laws may also make it difficult for women to hold a job and breastfeed.

The female breast is accentuated in books and magazines, by the media (especially television) and by manufacturers and advertisers of women's clothes. The breast may become regarded as a dominant sex symbol, and women may then not wish to breastfeed their babies in public, or they may falsely come to believe that breastfeeding will mar the appearance of the breasts. At the same time, a belief may develop that it is superior, chic and sophisticated to bottle-feed. Breastfeeding may be regarded as a primitive practice, and the feeding bottle may become a status symbol. As a result in many areas of the world breastfeeding is declining despite all recent efforts in its favour.

Traditional breastfeeding practices do not fit in with the demands of modern societies in which women have to be absent from their homes and their children for extended periods, usually for work. Although employment legislation in some countries provides for breastfeeding breaks for workers, distance from home and transport problems make it impractical for mothers to take advantage of the breaks. Thus, while it may well be possible for a mother to breastfeed her baby when they are together (usually at home), when they are apart the baby must be bottle-fed with infant formula. The mother could also express her own milk and leave it for someone else to feed to the baby using a bottle or a cup and spoon, but in practice few women do this. Some consider expressing milk a bother (although it is very easy once the technique is acquired) or unpleasant, and many worry about storing the breastmilk safely.

Giving babies breastmilk substitutes at an early age is dangerous even where breastfeeding is continued. The unnecessary very early partial replacement of breastmilk with breastmilk substitutes from a bottle introduces risks and sometimes serious problems for the infant, the mother and the family.

Management of Breastfeeding

If at all possible, breastfeeding should begin within minutes of delivery (or certainly within one hour). This early suckling has physiological advantages because it raises the levels of the hormone oxytocin secreted into the mother's blood. As described above, oxytocin causes uterine contractions which first help expel the placenta and then have an important role in reducing blood loss.

Soon after delivery the mother and her baby should be together in bed at home or in the hospital ward. In the past it was considered normal in modern hospitals to take the baby to a nursery ward and the mother to a maternity ward, but this practice is highly undesirable. Wherever "rooming-in" is not the current hospital practice, procedures need to be changed. It is absolutely safe for the baby to sleep in the same bed as the mother. There are very few contraindications (save serious illness in the mother or infant) for rooming-in or breastfeeding.

In the days after delivery, and as the baby gets older, breastfeeding should be done "on demand". That is, the baby should be breastfed when he or she wants to be fed and not,` as used to be common in Western countries, on a scheduled basis, such as every three or four hours.

The duration of feedings will vary and in general should not be limited.

Usually a baby feeds for 8 to 12 minutes, but there are fast and slow feeders, and both types usually get an adequate quantity of milk. Some mothers believe that the milk from the left breast is different from that from the right, but this is not so; the baby should feed from both breasts more or less equally.

Babies in the first few days after birth usually lose weight, so a baby with birth weight of 3 kg may weigh 2.75 kg at five days of age. A loss of up to about 10 percent is not unusual, but by seven to ten days the baby should have regained or overtaken the birth weight.

Almost all experts now agree that the infant should be exclusively breastfed for the first four to six months. An adequate gain in weight is the best way of judging the adequacy of the diet. No water, juices or other fluids are needed for a baby getting adequate breastmilk, even in hot humid or hot arid areas of the tropics; the baby will simply feed more often if thirsty. If the baby has diarrhoea breastfeeding should continue, but other fluids such as oral rehydration solutions or local preparations may be needed.

Experience from countries in East Africa, Asia and Latin America suggests that most mothers living in extended families in traditional societies are very successful, often very expert breastfeeders, and failure of lactation is uncommon. Traditional family life is undoubtedly of great importance to the beginning breastfeeder. Other women in the family provide the support and comfort - especially if there are difficulties - that people in Europe and North America have to seek from organizations such as La Leche League.

At clinics, time is often wasted on lessons on Western textbook ideas about breastfeeding, including insistence on burping, timing of feeds or frequent washing of the nipples. This emphasis on rules and strictures rather than on relaxation and pleasure is not good for anyone anywhere. It has been known to have grave psychological effects, often resulting in failure of lactation. The low rate of successful breastfeeders in North America and Western Europe is an indication of how inadequate Western-style breastfeeding has been, except in Scandinavia.

Breastfeeding should not be a complicated, difficult procedure. It should be enjoyable for both mother and child, and given the right circumstances of security, support and encouragement it can be. Some women in all societies do have problems with breastfeeding, but many of these are solvable or can be relieved. It is important that mothers have easy access to good advice and support. Many books dealing with lactation and related problems are available, and they should be consulted.

Common breastfeeding problems include:

— inverted or short nipples, or nipples that do not seem to be very protractile;
— nipples so long as to interfere with feeding, because some babies suck only the nipple and not the areola;
— refusal to feed, which needs to be checked in case the baby is ill or has a mouth problem such as a cleft palate;
— soreness of the breasts, which may be caused by cracked nipples, by mastitis or by a breast abscess requiring antibiotics and good medical care;
— so-called insufficient milk, which is discussed below;
— leakage from the breasts, which may cause embarrassment but is usually self-limiting and can be dealt with by expression of milk and by using an absorbent pad to prevent wetting of clothing.

Breastfeeding Problems

Complete lactation failure

Very few mothers - fewer than 3 percent-experience complete or nearly complete lactation failure. If the mother has serious difficulties, seeks help and really wants to breastfeed her very young baby, then some heroic methods may be necessary. The mother may need to be admitted to hospital and placed in a ward where other women are successfully breastfeeding. She and her infant should be examined for any physical reason for inability to breastfeed. The mother should be given plenty of fluids, including milk. These are mainly psychological inducements aimed to encourage lactation. In some societies local foods or potions are considered to be lactagogues, or substances that stimulate breastmilk production. There is no harm in trying these substances. A knowledgeable doctor or senior health worker may prescribe one of two drugs which are sometimes effective in improving or stimulating milk production: the tranquillizer chlorpromazine, 25 mg three times a day by mouth, or the newer drug metoclopramide, 10 mg three times a day.

In general, the important basis for treatment is to help the mother relax, to assist her in getting the baby to suck at the breast and to make certain that, while the breast is relied upon as much as possible, the baby is not losing weight. The dilemma is that the more the infant suckles at the breast, the greater the stimulation to production and let-down of milk; while the more the supplementary foods given, the less the infant will want to suckle.

If breastfeeding remains unsuccessful in an infant of up to three months of age, the mother should be taught to feed infant formula or milk to the baby either with a cup and spoon or with a feeding bowl. A cup and spoon are easier to keep clean than a bottle and teat. Some means should be found to provide the mother with adequate infant formula, fresh milk or full-cream milk powder if she cannot afford to buy it, which will often be the case. The infant should attend a clinic regularly.

This method of feeding also applies to the infant of a mother who dies in childbirth. It is then desirable to admit to hospital both the child and the female relative or care-giver who is to be responsible for the infant's feeding. An alternative is to find a lactating relative or friend to act as a wet nurse and to breastfeed the infant. Sometimes a friend or relative will be willing and able.

Failure of lactation or death of the mother after the infant is four months of age calls for a different regime. The child can be fed a thin gruel of whatever is the local staple food, to which should be added adequate quantities of milk or milk powder. It is advantageous to provide some extra fat in the infant's diet. A relatively small quantity of groundnut, sesame, cottonseed, red palm or other edible oil will markedly increase the baby's energy intake without adding too much bulk to the diet. If milk or milk powder is not available then any protein-rich food such as legumes, eggs, ground meat, fish or poultry may be used.

Insufficient milk production

Much more common than lactation failure is the belief by a mother that she is not producing enough breastmilk to satisfy her baby. Insufficient milk is very commonly reported by mothers in industrialized countries; perhaps the baby cries a lot or the mother feels that the baby is not growing adequately, or there may be any of a number of other reasons. In medicine this common condition is termed "insufficient milk syndrome". It is often at first a psychological concern rather than a serious condition, but it may rapidly lead to a real problem of milk production. Too often physicians, nurses and friends of the mother provide exactly the wrong advice to the mother concerned about her milk production.

In many studies, especially in industrialized countries, "insufficient milk" is cited as the most common reason given by mothers for their early termination of breastfeeding or for early supplementation with other foods, especially formula. It is all too easy to assume simply that many women are incapable of producing enough milk to feed their young infants. The busy practitioner's answer, when faced with a mother complaining of insufficient milk, is simply to advise her to supplement her breastmilk with bottle feeds. This may be exactly the wrong advice to give.

Suckling at the breast encourages the release of prolactin. The maintenance of lactation is dependent on adequate nipple stimulation by the suckling infant. It is now evident that diminishing breastmilk production results from reduced nipple stimulation. The cause of insufficient milk may therefore often be that alternative feeding has replaced breastfeeding to a variable degree. Therefore, advice to provide or increase supplementation is almost always going to contribute to a reduction in breastmilk production; supplementary bottle feeds are used as a cure for insufficient milk when in fact they are the cause.

The most appropriate treatment for insufficient milk syndrome in a mother who wishes to breastfeed is to advise her to try to increase milk production by putting the infant to the breast more frequently, in this way increasing stimulation of the nipples. The common medical advice, increasing bottle feeds, is likely to worsen the situation, leading to a further decline in milk production and eventual cessation of lactation. This is not to condemn supplementary feeding, especially after the infant is six months of age, but it should be clear that its use will almost inevitably contribute to a decline in milk production.

Maternal employment away from home is frequently cited as the most important reason for a decline in breastfeeding. Published surveys, however, seldom cite work as an important reason for not initiating breastfeeding or for early weaning from the breast. Clearly, employment out of the home for more than a few hours a day does place constraints on the opportunity to breastfeed and provides a reason for supplementary feeding. It may therefore contribute to the development of insufficient milk production.

Working mothers can continue to breastfeed successfully and can maintain good levels of lactation. Nipple stimulation from adequate suckling during the time they spend with their infants is particularly important for them. There is a need for labour laws and work conditions that recognize the special needs of lactating mothers in the labour force. If breastfeeding were accepted as necessary and usual practice by governments and employers, then arrangements would have to be made for a woman's baby to be near her for the first six months of life.

Past and present promotional practices by manufacturers of breastmilk substitutes may be an important factor contributing to the problem of insufficient milk. The companies find it advantageous to influence both the public and the medical profession to believe that supplementary bottle-feeding is the answer to insufficient milk.

The best and easiest way to judge whether or not a baby is getting enough breastmilk, when no other feeding is provided, is to weigh the baby regularly. Normal or near-normal weight gain provides the best evidence of adequate breastmilk production.

Breastfeeding, fertility and birth spacing

The traditional wisdom of many societies long included a belief that breastfeeding reduced the likelihood of an early pregnancy. Often this belief was regarded as an old wives' tale. Scientific evidence now proves beyond question that the intensity, frequency and duration of breastfeeding bears a positive relationship to the length of post-partum amenorrhoea, anovulation and reduced fertility. Mothers who breastfeed intensively find that there is a relatively long period after birth before menstruation resumes. In contrast, the interval between birth and the onset of monthly periods is short in women who do not breastfeed their babies. The physiology of this phenomenon is now reasonably clear; it is related to hormones produced as a result of sucking stimulation of the nipple.

This knowledge has important implications in terms of birth spacing and population dynamics. In many developing countries, breastfeeding now contributes more to child spacing and to prolonging intervals between births than does the combined use of contraceptive pills, intrauterine devices (IUDs), condoms, diaphragms and other modern contraceptives. Therefore the fertility-controlling benefits of breastfeeding should now be added to its other advantages.

Recent data from Kenya and elsewhere suggest that women who continue to breastfeed for a long time but also introduce bottle-feeding in the first few months of the infant's life may have shorter post-partum amenorrhoea than women who practise breastfeeding exclusively. The use of breastmilk substitutes in the first few months of life reduces sucking at the breast, thus lowering prolactin levels and leading to an earlier return of ovulation and menstruation even in mothers who breastfeed for a year or more. Thus bottle-feeding of babies contributes to a narrower spacing between births.

The so-called lactational amenorrhoea method (LAM) of natural family planning is now being widely and successfully used. If a mother has an infant under six months of age, is amenorrhoeic (with no vaginal bleeding from 56 days post partum) and is exclusively or almost fully breastfeeding her infant, then she is said to be 98 percent protected against pregnancy. She does not need to use any artificial family planning method.

Breastfeeding and AIDS

Human immunodeficiency virus (HIV) infection is now a major health challenge worldwide. Infection with HIV is followed, often some years later, by progressive disease and eventually by immunosuppression. The resulting syndrome, called acquired immunodeficiency syndrome (AIDS), is characterized by the development of various infections, often with diarrhoea and pneumonia, and malignancies such as Kaposi's sarcoma, leading eventually to death. In many developing countries HIV infection is almost as common in females as in males. Increasing numbers of infants and young children appear to be infected from their mothers. The exact mechanisms of transmission from the mother to the foetus or infant is not known. Transmission could occur in *utero* through passage of the virus across the placenta; around the time of delivery through exposure to vaginal secretions, ingestion of maternal blood or maternal-foetal transfusion during labour and delivery; and in infancy through ingestion of the virus in breastmilk. In many countries HIV infection has been reported in 25 to 45 percent of infants born to HIV-positive mothers.

Evidence suggests that HIV can be transmitted from infected mothers to their uninfected infants through breastmilk. The virus has been isolated from human breastmilk. It was thought that the fragile virus might be destroyed by gastric acid and enzymes in the infant's gut and that the stomach and intestines of infants might be relatively impervious to the virus. This is probably largely true; by far the majority of babies breastfed by HIV-infected mothers do not become infected through breastmilk. It has been difficult, however, to determine whether a particular infant was infected prior to delivery, at the time of delivery or through breastfeeding. This uncertainty is partly due to the fact that both infected and uninfected infants acquire HIV antibodies passively from their infected mothers, but the presence of antibodies in standard HIV tests cannot be interpreted to mean active infection.

A pregnant woman with poor vitamin A status is more likely than others to pass the HIV infection to the foetus. Transmission from mother to infant through breastmilk is currently thought to be relatively rare. Some apparent differences in rates of transmission among groups of women from different countries may be related to vitamin A intake and other factors.

Many infants in Africa, Asia and Latin America live in settings where gastrointestinal infections are prevalent, hygiene is poor and water supplies are suspect. In these circumstances the many advantages of breastfeeding far outweigh the risk to the infant of AIDS infection through breastmilk from an HIV-positive mother. Only where the common causes of morbidity and mortality in infancy are not infectious diseases should public policy advise the use of bottle-feeding in place of breastfeeding to reduce the possibility of AIDS transmission. The individual mother should, of course, where feasible, be counselled by a doctor or trained health worker and cautioned about the relative risks to the infant of breastfeeding or alternative feeding methods in terms of disease and survival. This counselling will allow the mother to make an informed decision.

References

Brown, M.L. (1990). *Present knowledge in nutrition.* Washington, DC, USA, International Life Sciences Institute, Nutrition Foundation. 6th ed.

Cameron, M. & Hofvander, Y.(1983). *Manual on feeding infants and young children.* Oxford, UK, Oxford University Press. 3rd ed.

King, F.S. & Burgess, A.(1993). *Nutrition for developing countries.* Oxford, UK, Oxford University Press. 2nd ed.

United States Department of Agriculture (USDA). (1976-88). *Composition of foods.* Agriculture Handbooks Nos.1,4,9,11, 16. Washington, DC, USA.

WHO.(1993). *Breast-feeding. The technical basis and recommendations for action.* Geneva, Switzerland.

Bibliography

Asp, N.-G. (1995). Classification and methodology of food carbohydrates as related to nutritional effects. *American Journal of Clinical Nutrition* 61(4(S)):930S -937S.

Barker, Helen M. (2002). *Nutrition and dietetics for health care*. Edinburgh: Churchill Livingstone. p. 17.

Berg J, Tymoczko JL, Stryer L (2002). *Biochemistry* (5th ed.). San Francisco: W.H. Freeman. p. 603.

Brown, M.L. (1990). *Present knowledge in nutrition.* Washington, DC, USA, International Life Sciences Institute, Nutrition Foundation. 6th ed.

Brun, T.A. & Latham, M.C. (1990). *Maldevelopment and malnutrition.* World Food Issues, Vol. 2. Ithaca, New York, Program in International Agriculture.

Cameron, M. & Hofvander, Y.(1983). *Manual on feeding infants and young children.* Oxford, UK, Oxford University Press. 3rd ed.

Carroll, K.K. and Khor, H.T. (1975). Dietary fat in relation to tumorigenesis. *Progress in Biochemical Pharmacology.* 10: 308-353.

Commission on Life Sciences. (1985). *Nutrition Education in US Medical Schools*, p. 4. National Academies Press.

Corbridge D. E. C. (1995). *Phosphorus: An Outline of its Chemistry, Biochemistry, and Technology* (5th ed.). Amsterdam: Elsevier.

Curley, S., and Mark (1990). *The Natural Guide to Good Health*, Lafayette, Louisiana, Supreme Publishing

Department of Health. (1989). *Dietary sugars and human health.* Her Majesty's Stationery Office, London.

Englyst, H.N. and Hudson, G.J. (1996). The classification and measurement of dietary carbohydrates. *Food Chemistry,* 57(1): 15-21.

Eveleth, P.B. & Tanner, J.M. 1976. *Worldwide variation in human growth.* Cambridge, Cambridge University Press.

FAO. (1980). Carbohydrates in human nutrition, a Joint FAO/WHO Report. *FAO Food and Nutrition Paper 15,* Rome.

Ferro-Luzzi, A. et al. 1979. Nutrition, environment and physical performance of preschool children in Italy. In: Somogyi, C. & de Wijn, J.F., ed. *Nutritional aspects of physical performance.* Basel, Karger, pp. 85–106.

Galdston, I. (1960). *Human Nutrition Historic and Scientific.* New York: International Universities Press.

Garn, S.M. In: Farmer, F.A., ed. 1978. *Nutrition of the aged.* Alberta, University of Calgary, pp. 73–90.

King, F.S. & Burgess, A.(1993). *Nutrition for developing countries.* Oxford, UK, Oxford University Press. 2nd ed.

Lippard, S. J. and Berg, J. M. (1994). *Principles of Bioinorganic Chemistry.* Mill Valley, CA: University Science Books.

Mahan, L.K. and Escott-Stump, S. eds. (2000). *Krause's Food, Nutrition, and Diet Therapy* (10th ed.). Philadelphia: W.B. Saunders Harcourt Brace.

Nelson, D. L.; Cox, M. M. (2000). *Lehninger Principles of Biochemistry* (3rd ed.). New York: Worth Publishing.

Paøízková, J. In: Chandra, R.K., ed. 1981. *Critical reviews of tropical medicine,* Vol. 1. New York, Plenum Publ. Corp.

Reed, D., McGee, D., Yano, K. and Hankin, J. (1985). Diet, blood pressure and multicollinearity. *Hypertension.* 7: 405-410.

Rolls, B.J. and Shide, DJ. (1992). The influence of dietary fat on food intake and body weight. *Nutrition Reviews.* 50: 283290.

Shils et al. (2005). *Modern Nutrition in Health and Disease.* Lippincott Williams and Wilkins.

Southgate, D.A.T. (1991). *Determination of food carbohydrates.* Elsevier Science Publishers, Ltd., Barking.

Thiollet, J.-P. (2001). *Vitamines & minéraux.* Paris: Anagramme.

UN ACC/SCN.(1989). *Malnutrition and infection. A review,* by A. Tomkin & F. Watson. Geneva, Switzerland.

United States Department of Agriculture (USDA). (1976-88). *Composition of foods.* Agriculture Handbooks Nos.1,4,9,11, 16. Washington, DC, USA.

Walter C. Willett and Meir J. Stampfer (January 2003). "Rebuilding the Food Pyramid". *Scientific American* 288 (1): 64–71.

Waterlow, J.C. (1992). *Protein energy malnutrition.* London, UK, Edward Arnold.

WHO Technical Report Series. Diet, nutrition and the prevention of chronic diseases." Report of a Joint WHO/FAO Expert Consultation; Geneva 2003. Retrieved 2011-03-07

World Bank.(1994). *Enriching lives. Overcoming vitamin and mineral malnutrition in developing countries.* Washington, DC, USA.

World Health Organization. 1983.*Measuring change in nutritional status.* Geneva, World Health Organization.